THE

YOUNG CHEMIST:

A BOOK

OF

LABORATORY WORK,

FOR BEGINNERS.

BY

JOHN HOWARD APPLETON, A. M.,

PROFESSOR OF CHEMISTRY IN BROWN UNIVERSITY.

EIGHTH EDITION.

SILVER, BURDETT & COMPANY

NEW YORK ... BOSTON ... CHICAGO

PROFESSOR APPLETON'S

SERIES OF CHEMICAL TEXT-BOOKS:

I. The Beginner's Handbook of Chemistry: *Price*, $1.00. This is an introduction to the study of Chemistry, suitable for general readers. It treats chiefly the non metals, these being generally found to furnish the best material for an elementary course, and to best illustrate the fundamental facts and principles of the science.

The book is written in an attractive style, and has had a very large sale. It is profusely illustrated with engraving, and has, in addition, fourteen colored plates.

II. The Young Chemist: *Introductory Price*, 75 *Cents*. A book of chemical experiments for beginners in Chemistry. This is designed for use in Schools and Colleges. It is composed almost entirely of experiments, those being chosen that may be performed with very simple apparatus. The book is arranged in a clear, systematic, and instructive manner.

III. Report-book of Chemical Experiments. *First Series. Introductory Price*, 25 *Cents*. A well arranged memorandum-book, with blank spaces to be filled by the pupil during the progress of his experiments.

The making of a succinct report by the student, is of great service in leading him to form the habit of taking written notes while the facts of the experiment are fresh in the mind. Morover, it undoubtedly increases the powers of observation.

This Report-book is so constructed that it may be used with "The Young Chemist," or with any text-book on general chemistry.

IV. Qualitative Analysis: *Introductory Price*, 75 *Cents*. A brief but thorough manual for laboratory use.

It gives full explanations and many chemical equations. The processes of analysis are clearly stated, and the whole subject is handled in a manner that has been highly commended by a multitude of successful teachers of this branch.

V. Quantitative Analysis: *Introductory Price*, $1.25. A text-book for school and college laboratories.

This volume possesses novel and striking merits such as will make it worthy of the same decided approbation and large sale that have been awarded to the earlier books of this series. The treatment of the subject is such that the pupil gains an acquaintance with the best methods of determining all the principal elements, as well as with the most important type-processes both of gravimetric and volumetric analysis.

THE EXPLANATIONS ARE DIRECT AND CLEAR, so that the pupil is enabled to work intelligently *even without the constant guidance of the teacher.* By this means the book is adapted for self-instruction of teachers and others who require this kind of help to enable them to advance beyond their present attainments.

VI. Chemical Philosophy: *Introductory Price*, $1.40. A text-book for schools and colleges.

It deals with certain general principles of chemical science, such as the constitution of matter; atoms, molecules, and masses; the three states of matter and radiant matter, the change of state from one form of matter to another. It also presents such topics as Boyle's and Mariotte's law, Charles' law, and the other general laws of matter. It discusses from a chemical standpoint certain forms of energy, such as heat, light, electricity. It treats of the nature of chemical affinity; the chemical work of micro-organisms; the modes of chemical action; thermo-chemistry; and those attractions of substances which are partly physical and partly chemical. It also presents a full study of atomic weights: the methods leading to a first adoption of them, and then to the grounds sustaining certain numbers selected. The periodic system is of course discussed.

The work is fully illustrated.

Copies sent by mail, postpaid, *upon receipt of Introductory price,* by

SILVER, BURDETT & CO., 219-223 Columbus Ave., Boston.

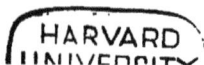

PREFACE
TO THE REVISED EDITION.

THE purpose of this little book is to aid in the instruction of pupils in chemistry. The method employed is the experimental or object method.

Every experienced teacher has remarked the wonderful ease and pleasure with which beginners in chemistry—*when they are allowed to perform experiments*—grasp the facts and principles of the science. It has also been recognized that the only objections to the experimental method arise from the greater expenditure of the teacher's time, and from the cost of supplies.

It is hoped that this little book removes one of these objections; and, fortunately, chemical apparatus and supplies can now be had at very low prices.

The following are some of the characteristic advantages of the book—

First.—The apparatus described, and the supplies called for, are of *the very simplest character.*

Second.—The experiments are described in clear and simple language, and in direct form; the pupil can hardly fail to perform them successfully, even without special aid from the teacher.

Third.—Dangerous experiments have been excluded. (But, of course, care must always be exercised in experimenting.)

Fourth.—The chemical elements are discussed in a scientific order which, while it aids the memory, does so upon correct principles.

Fifth.—Formulas and reactions are introduced freely, so that the student learns the new nomenclature and new notation without suspecting it. (But a systematic discussion of these subjects has been offered for purposes of reference, or for such other use as the teacher may judge best to make of it.)

It may also be added that this book is not an experiment. For many years it has been used with great success by many professors and teachers of wise judgment and large experience.

The present edition has been carefully revised throughout, and it is hoped that in its improved form it may be found to possess additional usefulness.

BROWN UNIVERSITY, 1892.

CONTENTS.

PAGE

HINTS TO TEACHERS.. 7

INTRODUCTION.

Nomenclature and Notation of Chemistry............................. 11

   First Section.—Elements and Compounds..................... 11
   Second Section.—Names and Symbols and Formulas........... 14
   Third Section.—Systematic Names of Compounds............. 17

CHAPTER I.—THE NON-METALLIC MONADS............ 25
   Hydrogen 26
   Fluorine. 29
   Chlorine... 30
      Hydrochloric acid. 32
   Bromine. 33
   Iodine 34

CHAPTER II.—THE NON-METALLIC DYADS.............. 37
   Oxygen.. 38
   Sulphur... 42
      Sulphuric acid... 43
      Sulphuretted-hydrogen. 45
   Selenium and Tellurium.. 45

CHAPTER III.—THE NON-METALLIC TRIADS............ 46
   Boron 47
      Boric acid... 47
   Nitrogen 48
      Compounds of Nitrogen and Hydrogen...................... 48
         Ammonia-gas.. 49
      Compounds of Nitrogen and Oxygen........................ 50
         Nitrogen dioxide...................................... 51
         Nitrogen pentoxide.................................... 51
         Nitric acid... 51
   Phosphorus.. 54
   Arsenic 55
   Antimony 57

PAGE

CHAPTER IV.—THE **NON-METALLIC TETRADS**........ 59
  Carbon.. 60
    Compounds of Carbon and Hydrogen 61
      Ethylene.. 62
    Compounds of Carbon, Hydrogen, and Oxygen................ 62
    Compounds of Carbon and Oxygen........................... 62
      Carbon monoxide.. 62
      Carbon dioxide........ 63
  Silicon... 64
  Titanium.. 65
  Tin 66

CHAPTER V.—THE **METALLIC MONADS**.... 67
  Silver... 68
  Potassium 69
  Sodium 71
  Lithium.. 71

CHAPTER VI.—THE **METALLIC DYADS**...................... 72
First Section.
  Lead... 73
  Barium.. 75
  Strontium... 76
  Calcium... 77

Second Section.
  Mercury... 80
  Copper.. 82
  Magnesium... 84
  Zinc 86

Third Section.
  Cobalt 89
  Nickel.. 90
  Iron 91
  Manganese... 93
  Chromium.. 94
  (Aluminium)... 96

CHAPTER VII.—THE **METALLIC TRIADS**.................... 98
  Bismuth 99
  Gold... 100

CHAPTER VIII.—THE **METALLIC TETRAD** 101
  Platinum.. 101

APPENDIX.

List of Chemical Supplies Needed.......................... 103
List of Apparatus Needed 106

Hints to Teachers.

I. Perform slowly several experiments before the class. Let the pupils perform the same experiments (and no others), each at his own desk. After this let the pupils learn carefully the entire description of the experiments so performed.

It is highly desirable to have the pupils learn the *outline* of a given chapter, and recite it day after day, until the work of that chapter is finished. They thus discover the logical relation which binds the separate experiments into one whole; they also discover the scientific plan of the work.

II. Use extreme caution in experimenting. Be careful not to vary the conditions of an experiment, as stated in the book. Be exceedingly careful when you attempt experiments other than those described in this book.

Do not allow pupils to approach too near to an experiment in progress.

III. Use very small quantities of the substances prescribed.

IV. In preparing a gas, the most convenient apparatus is a side-neck flask or a side-neck test-tube.

The cuts need no explanation.

Fig. 1.—Evolving a gas by use of a side-neck flask.

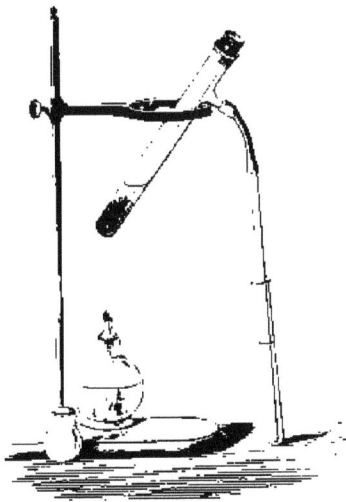

Fig. 2.—Evolving a gas by use of a side-neck test-tube.

7

V. To collect a gas in a small bell-glass, use a *lead-post*. This is made by cutting a strip of lead into the form shown at the left, in Fig. 3, and folding it into the other form shown in the same figure.

Fig. 3.—The lead-post before bending into shape, and after bending.

The use of a rubber ring in attaching a test-tube to the lead-post is apparent upon inspection of Fig. 4.

Fig. 4.—Illustrating the preparation of gas by use of the lead-post, the side-neck test-tube, and the wire triangle.

VI. If more than one experiment is to be performed with a given gas, several portions of gas may be collected in several small bottles; the gas may be retained a short time in the bottles by covering the latter, when filled, with wet pieces of filter-paper.

VII. Instead of being placed *alongside* of a beaker or casserole, the lead-post may be placed *inside* of a water-pan of granite-ironware, or other suitable ware.

VIII. As a support for apparatus, a wire triangle arranged on screw-eyes as in Fig. 4 is very useful and very cheap. The teacher's own ingenuity will suggest a variety of modifications of this triangle, so as to suit a variety of purposes.

IX. As a support or prop for lamps, etc., wooden blocks, from three to four inches square and from one-half to one inch thick, are extremely serviceable. In Fig. 1 both thicknesses are represented.

HINTS TO PUPILS.

I. Follow very closely the directions under each experiment.

II. Be *exceedingly careful* when you try any experiment not described in this book.

III. Do not have the face, and especially the eyes, too near an experiment in progress.

IV. Be *especially cautious*—

    (*a*) When heating substances;

    (*b*) When using sulphuric acid;

    (*c*) When using phosphorus;

    (*d*) When using combustible liquids like alcohol, or, indeed, anything that can burn.

INTRODUCTION.

THE NOMENCLATURE AND NOTATION OF CHEMISTRY.

OUTLINE OF THIS CHAPTER.

FIRST SECTION.—*Elements and Compounds.*

An Element.—A Compound.—A mechanical mixture.—List of Elements with their atomic weights.

SECOND SECTION.—*Names, Symbols, and Formulas.*

Names of Elements.—Literal, graphic, and glyptic symbols of Elements.

Names of Compounds.—Literal, graphic, and glyptic formulas of Compounds.

THIRD SECTION.—*Systematic Names of Compounds.*

1st. Names of Binaries. — Compound Radicles. — Anhydrides. — Haloid acids.—Haloid salts.

2d. Names of Ternaries.—Acids.—Salts (normal, acid, and basic).–Graphic symbols of ternary acids and salts.

FIRST SECTION.

ELEMENTS AND COMPOUNDS.

1. An **element** or elementary substance is a form or kind of matter that cannot, by any *known* means, be decomposed or subdivided into parts differing from itself.

11

For example, sulphur cannot be decomposed, by any *known* means,
into parts differing from sulphur.

Also, pure iron cannot be decomposed, by any *known* means, into parts
differing from iron.

Sulphur is an element; iron is an element.

2. A **compound** is formed by the chemical union of
elements. A compound may be broken up or decomposed, *by chemical means*, into the elements of which it is
composed. But a compound cannot be decomposed, by
mere mechanical subdivision, into its elements.

For example, sulphur (**S**) and iron (**Fe**) may form a chemical union.
The product is a chemical compound, called ferrous sulphide, and indicated by the formula, **FeS**. This compound may be decomposed chemically into iron and sulphur; but by no mere mechanical means can we
take away the one element from the other, when they are combined chemically into a compound. Moreover, the compound, formed by iron and sulphur, is very different in most of its properties from iron and from sulphur.

3. A **mechanical mixture** is formed when two substances are merely intermingled, without chemical union.

For example, filings of iron and powdered sulphur may be intermingled
to form a mechanical mixture. But, by means of a sieve of proper fine-
ness, the sulphur may be entirely sifted out from the iron filings.

4. No complete list of *mechanical mixtures* can be
given. The number of such possible mixtures appears
to be infinite.

5. No complete list of *chemical compounds* can be
given. We do not know that there is any limit to the
number of them. A list of the known chemical compounds would be very large.

6. The *chemical elements* now known are far less numerous. The total number is about seventy. About 99
per cent. of the matter of our planet is made up of only
thirteen of the elements, united in various compounds.
The other elements exist in relatively small quantities.

The Chemist's Elementary Substances.

Name of Element.	Atomic Symbol.	Approximate Atomic Weight.	Name of Element.	Atomic Symbol.	Approximate Atomic Weight.
Aluminium .	Al	27.	Molybdenum.	Mo	95.5
Antimony . .	Sb (Stibium) . . .	120.	Nickel . . .	Ni	57.9
Arsenic . . .	As	74.9	Niobium . .	Nb	93.8
Barium . . .	Ba	136.8	Nitrogen . .	N	14.
Bismuth . .	Bi	207.5	Osmium . .	Os	198.5
Boron	B	10.9	Oxygen . . .	O	16.
Bromine . .	Br	79.8	Palladium . .	Pd	105.7
Cadmium . .	Cd	111.8	Phosphorus .	P	31.
Cæsium . . .	Cs	132.6	Platinum . .	Pt	194.4
Calcium . . .	Ca	40.	Potassium. .	K (Kalium)	39.
Carbon . . .	C	12.	Rhodium . .	Rh	104.1
Cerium . . .	Ce	140.4	Rubidium . .	Rb	85.3
Chlorine . .	Cl	35.4	Ruthenium .	Ru	104.2
Chromium . .	Cr	52.	Samarium . .	Sm	150.
Cobalt . . .	Co	58.9	Scandium . .	Sc	44.
Copper . . .	Cu (Cuprum) . . .	63.2	Selenium . .	Se	78.8
Didymium . .	D	144.6	Silicon . . .	Si	28.2
Erbium . . .	E	165.9	Silver . . .	Ag (Argentum) . .	107.7
Fluorine. . .	F	19.	Sodium . . .	Na (Natrium). . .	23.
Gallium . . .	Ga	68.9	Strontium . .	Sr	87.4
Germanium .	Ge	72.3	Sulphur . . .	S	32.
Glucinum . .	G or Be (Beryllium).	9.1	Tantalum . .	Ta	182.1
Gold	Au (Aurum	196.2	Tellurium . .	Te	128.
Hydrogen . .	H	1.	Thallium . .	Tl	203 7
Indium . . .	In	113.4	Thorium . .	Th	233.4
Iodine. . . .	I	126.6	Tin	Sn (Stannum) . . .	117.7
Iridium . .	Ir	192.7	Titanium . .	Ti	48.
Iron	Fe (Ferrum)	55.9	Tungsten . .	W (Wolframium) .	183.6
Lanthanum .	La	138.5	Uranium . .	U	238.5
Lead	Pb (Plumbum) . .	206.5	Vanadium . .	Va	51.3
Lithium . . .	Li	7.	Ytterbium . .	Yb	172.8
Magnesium .	Mg	24.	Yttrium . . .	Y	89.8
Manganese .	Mn	53.9	Zinc	Zn	64.9
Mercury . . .	Hg (Hydrargyrum)	199.7	Zirconium . .	Zr	89.4

SECOND SECTION

CHEMICAL NOMENCLATURE AND NOTATION.

7. A CHEMICAL substance may be designated by a name, or it may be represented more in brief by a symbol or formula.

The same substance may properly have more than one name, and it may be correctly represented by more than one symbol or formula.

8. A system of chemical nomenclature and notation aims to employ names, symbols, and formulas which shall represent the true qualitative and quantitative composition of substances.

Names of Elements.

9. No special system is necessary in the case of elements, but it is customary, (*a*) to allow the names of elements long known, to remain unchanged—*e. g.*, Gold; (*b*) to derive the names of new elements from some well-marked property of them—*e. g.*, Chlorine, a greenish gas, derives its name from a Greek word (χλωρός, *chloros*) meaning light green; (*c*) the names of newly-discovered *metals* are made to terminate in *um*—*e. g.*, Thallium.

Symbols of Elements.

10. Literal symbols are those which employ *letters*. An atom of an elementary substance is usually indicated by the initial (sometimes with the addition of another letter) of its native or of its Latin name, thus:

C indicates one atom of Carbon;
Ca   "   "   Calcium;
Cd   "   "   Cadmium;

Ce indicates one atom of Cerium;
Cl " " Chlorine;
Co " " Cobalt;
Cr " " Chromium;
Cs " " Cæsium;
Cu " " Copper (cuprum).

11. Graphic symbols are those which employ *dia-grams.* Thus, Professor Kekulé recommends the following symbols—

—to represent monad, dyad, triad, and tetrad atoms or radicles respectively.

The same symbols may be conveniently simplified to the following forms :

12. Glyptic symbols are those which employ *models,* as spheres, cubes, tetrahedrons, etc. Sometimes models having different colors are used, so as to suggest the properties of the substances represented.

Names of Compounds.

13. Most chemical compounds have more than one name. Sometimes the same compound has three or four different names.

There may be—

(*a*) A name *strictly descriptive of the components;* thus, the compound of Hydrogen and Chlorine (HCl) is called Hydrogen chloride ;

(*b*) A name *suggestive of some property* of the substance : thus the compound above mentioned (HCl) is called Hydrochloric acid or Chlorohydric acid ;

(*c*) A *commercial* name; thus, HCl is called, in commerce, Muriatic acid;

(*d*) A *mineralogical* name; thus, the compound of
Lead and Sulphur (PbS), is called, properly, Lead sulphide; but the mineral substance, found crystallized in
nature, and having the composition PbS, is called Galena;

(*e*) A more or less *arbitrary* name. This is exem-
plified in the case of many organic compound radicles;
thus, the compound of carbon and hydrogen having the
constitution represented by the symbol H_4C is usually
called Marsh-gas.

Formulas of Compounds.

14. Literal Formulas.—The literal formula of a compound is formed by grouping together the literal symbols
of the elements composing it. It is customary to place
the symbol of the most electro-positive substance first,
and in general to arrange the symbols so as to follow the
order of the parts of the name of the compound. But,
where no special effort is made to indicate the arrange-
ment of atoms in the molecule, the formula is said to be
empirical; thus, HNO_3 is an empirical formula for Nitric
acid. Where such attempt is made, the formula is called
rational; thus the rational formula of Nitric acid is
$H—O—(N\equiv O_2)$.

15. Graphic Formulas.—Of course, all graphic formulas are rational formulas; they are also *general* formulas.

As examples of the Kekulé system, Nitric acid, HNO_3
is represented thus:

It may also be represented thus:

Water, H_2O, thus, ⊙⊙; Mercuric chloride, $HgCl_2$,

thus, ⊞; Mercurous chloride, Hg_2Cl_2, thus, ⊞.

16. Glyptic Formulas.—These are models, made by joining together the glyptic symbols of elements.

THIRD SECTION.

STRICTLY SYSTEMATIC NAMES OF COMPOUNDS.

1st. Binaries.

17. Definition.—A binary is a compound which has but two kinds of atoms.

Thus,   HCl, Hydrogen chloride, is a binary;

SO_3, Sulphuric oxide, is a binary.

18. Compound Radicles.—Sometimes the term binary is extended to apply to a union of two compounds, called compound radicles, which play the parts of two elements.

Thus (NH_4), a compound radicle called Ammonium, and (CN), a compound radicle called Cyanogen, may unite to form the compound $(NH_4)(CN)$, called Ammonium cyanide, which may be considered a binary.

19. Names.—In case of binaries, the name given *involves* the names of both parts of the binary. The termination of the second name (which is always that of the more electro-negative substance) is always changed to *ide*. The termination of the first name (which is always that of the more electro-positive substance) is changed to *ous* or *ic*, or else remains unchanged, according to its equivalence. When no attempt is made

2 *

B

to express the idea of equivalence the name is unchanged; thus, AgCl is silver chloride. But *ous* is usually employed for lower, and *ic* for higher, equivalences:

Thus, $\overset{IV}{SO_2}$ is Sulphurous oxide; and $\overset{VI}{SO_3}$ is Sulphuric oxide.

20. Prefixes.—Prefixes are sometimes used. They may
be *numeral*, as Manganese di-oxide for MnO_2; or they
may be *general;* thus, *hypo* is used for a lower, and *per*
(abbreviation for hyper) is used for a higher, equivalence.

21. Anhydrides.—*An acid anhydride* is a substance—
usually a binary—which, by combining with water, or
some analogous compound, can produce a ternary called
an acid. An acid anhydride is usually composed of a
non-metal united with oxygen.

Thus, SO_3, Sulphuric oxide, is also called Sulphuric anhydride, because
it can combine with water to form a ternary acid—H_2SO_4, Sulphuric acid.

Again, SO_2, Sulphurous oxide, is also called Sulphurous anhydride, because
it can combine with water to form a ternary acid—H_2SO_3, Sulphurous acid.

A basic anhydride is a substance—usually a binary—
which, by combining with water or some analogous compound, can produce a ternary called a base or hydroxide.
The basic anhydride is usually composed of a *metal*
united with oxygen.

Thus, CaO, Calcium oxide, is also called a basic anhydride because it
can combine with water to form a base—CaO_2H_2, Calcium hydroxide.

22. Haloid Acids.—Though most acids are ternaries,
there are some acids that are binaries; as, HCl, Hydrochloric acid. Such acids are called haloid acids.

23. Haloid Salts.—There is an important class of
salts, called *haloid* salts, the members of which are *binaries*.
They are formed after the analogy of common salt, NaCl.
KI, Potassium iodide, and KBr, Potassium bromide, are
examples. They are formed by the substitution of a metal
or radicle for the Hydrogen of certain corresponding
Haloid acids, such as HCl, Hydrochloric acid, and HI,
Hydriodic acid.

2d. Ternaries.

24. Definition.—A ternary is a compound of three parts; the first and third parts may each be represented, according to circumstances, either by single atoms, or by groups of atoms—or by compound radicles—without any peculiar restriction as to equivalences. The second part is the linking part, whence it cannot be a *monad;* it is oftenest one or more atoms of Oxygen.

The principal ternary compounds are acids and salts.

25. Acids.—An acid is a compound of Hydrogen, such that the Hydrogen may be removed, and a metal or metals, a radicle or radicles, may be substituted in its place in such a way as to give rise to a metallic salt.

The general formula for an acid is $H-D-\bar{R}$; in which H represents Hydrogen; D represents the linking dyad, usually Oxygen; $\bar{R}$ represents an electro-negative radicle (either simple or compound).

The following set of acid anhydrides may be used to illustrate the foregoing definition:

Cl_2O,	Hypochlorous anhydride;
Cl_2O_3,	Chlorous anhydride;
Cl_2O_5,	Chloric anhydride;
Cl_2O_7,	Perchloric anhydride.

The following reactions illustrate the system both of forming and of naming acids:

The acid anhydrides react with water as follows:

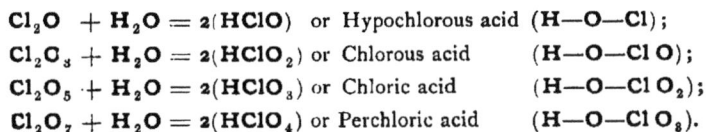

$Cl_2O \ + H_2O = 2(HClO)$ or Hypochlorous acid $(H-O-Cl)$;

$Cl_2O_3 + H_2O = 2(HClO_2)$ or Chlorous acid $\quad (H-O-Cl\,O)$;

$Cl_2O_5 + H_2O = 2(HClO_3)$ or Chloric acid $\quad (H-O-Cl\,O_2)$;

$Cl_2O_7 + H_2O = 2(HClO_4)$ or Perchloric acid $\quad (H-O-Cl\,O_3)$.

26. Salts.—A salt is a ternary linked by a dyad. The general formula of a salt is $\overset{+}{R}$—D—$\overset{-}{R}$; in which $\overset{+}{R}$ represents an electro-positive radicle (either simple or compound); D represents the linking dyad, usually Oxygen (and it should be remembered that there is usually one atom of linking dyad for each open point of attraction of the metal or positive radicle); $\overset{-}{R}$ represents an electro-negative radicle, which may be either simple or compound, but is usually made up of a non-metal combined with saturating oxygen (or with whatever dyad may be performing the linking function).

27. Salts may be viewed as formed by substitution of a metal, or other electro-positive radicle, for the Hydrogen of the acid from which the salt is formed.

Thus, Potassium may be substituted for the Hydrogen in the above acids, and may give rise to the following salts :

K Cl O,	Potassium hypochlorite,	(**K—O—Cl**);
K Cl O$_2$,	Potassium chlorite,	(**K—O—Cl O**);
K Cl O$_3$,	Potassium chlorate,	(**K—O—Cl O$_2$**);
K Cl O$_4$,	Potassium perchlorate,	(**K—O—Cl O$_3$**).

28. From the foregoing examples it will be seen that in naming a salt, the names of only two of the constituents are usually involved. The third constituent is so often Oxygen that the name of this element is *understood.* But, if the linking dyad is Sulphur, its name is expressed. The two constituents, whose names are always expressed, are the metal, and the non-metal which is the basis of the compound radicle. The Latin name of the metal is often used, and it is made to terminate in *ic* for higher and in *ous* for lower equivalences ; the name of the non-metal is made to terminate in *ate* when the salt is formed from an *ic* acid, or in *ite* when the salt is formed from an *ous* acid.

Thus, Ferrous sulphate ($FeSO_4$) is formed from an *ic* acid—that is, H_2SO_4, or Sulphuric acid.

Ferrous sulphite ($FeSO_3$) is formed from an *ous* acid—that is, H_2SO_3, or Sulphurous acid.

The following formulas illustrate the analogy of Sulphur salts to ordinary Oxygen salts:

H_3 As O_3 is Arsenious acid.

H_3 As O_4 is Arsenic acid.

K_3 As O_4 is Potassium arsenate, or, in full, Potassium oxy-arsenate.

K_3 As S_4 is Potassium sulpho-arsenate.

Salts may be *acid, normal,* or *basic.*

29. Normal salts.—They are called *normal salts* when all the Hydrogen, of the original acid, is replaced. K_2SO_4, Potassium sulphate, is a normal salt.

30. Acid salts.—They are called *acid salts* when only a part of the Hydrogen, of the original acid, is replaced —*e. g.,* Hydro-potassium sulphate, HK_2SO_4, formed from H_2SO_4. *Acid salts are part acid and part salt.*

31. Basic salts.—They are called *basic salts*, after the analogy of the term *base*, which was formerly applied to hydrates, such as PbO_2H_2, Lead hydroxide, $Pb{-O-H \atop -O-H}$.

Now, when the radicle NO_2 is substituted for both atoms of H, we have $Pb{-O-NO_2 \atop -O-NO_2}$, or $Pb(NO_3)_2$, which is the *normal* Lead nitrate. When the radicle, NO_2, is substituted for only one atom of Hydrogen, we have the product $Pb{-O-NO_2 \atop -O-H}$, or $Pb(NO_3HO)$, a *basic* salt (Lead nitro-hydrate). Basic salts are part base, and part salt.

The following table will assist in displaying the relationships of acids, bases, acid salts, basic salts, and normal salts to one another:

Acid,	$H{-O- \atop -O-}SO_2.$	Base,	$Pb{-O-H \atop -O-H}.$
Acid salt,	$H{-O- \atop K-O-}SO_2.$	Basic salt,	$Pb{-O-H \atop -O-NO_2}.$
Normal salt,	$K{-O- \atop K-O-}SO_2.$	Normal salt,	$Pb{-O-NO_2 \atop -O-NO_2}.$

Formulas of Acids and Salts.

32. Literal Formulas.—The manner of constructing *literal* formulas is apparent from the foregoing discussion.

33. Graphic Formulas.—A simple and useful method of representing acids is as follows:

HNO_3	H_2SO_4	H_3PO_4	H_4SiO_4

Of course these diagrams represent, in general, acids having, respectively, one, two, three, four atoms of replaceable hydrogen, and one, two, three, four atoms of linking oxygen, and attached to suitable electro-negative radicles.

They also represent, in general, the appropriate salts formed from the acids mentioned—the only restriction being that in the four examples given in the above paragraph the electro-positive constituents *must be monads.* But, of course, positive elements or radicles of higher equivalences may be indicated by using proper symbols.

Thus, the diagrams on the opposite page represent—by the simple combination of symbols similar to those indicated in paragraphs 15 and 33—a large number of the possible salts formed by such acids with monad, dyad, triad, and tetrad metals or positive radicles.

34. In connection with page 23, it may be said that in drawing diagrams it is desirable to employ continually the same plan. The following principles are recommended. Let the diagrams of ternary salts take the form of the letter **L**, so far as is practicable; let the linking dyad be always represented by vertical (or up-and-down) strokes; let the acid radicle be represented by horizontal (or right-and-left) strokes; let the metals or positive radicles be represented at the top.

These diagrams assist the student to comprehend and to remember formulas; and they cannot be expected to do more.

	Acid Radicles, with One Atom of Linking Oxygen.	Acid Radicles, with Two Atoms of Linking Oxygen.	Acid Radicles, with Three Atoms of Linking Oxygen.	Acid Radicles, with Four Atoms of Linking Oxygen.
With Monad Metals.	KNO_3	K_2SO_4	K_3PO_4	K_4SiO_4
With Dyad Metals.	$Pb(NO_3)_2$	$PbSO_4$	$Ca_3(PO_4)_2$	$Ca_2(SiO_4)$
With Triad Metals.	$Bi(NO_3)_3$	$Bi_2(SO_4)_3$	$BiPO_4$	$Bi_4(SiO_4)_3$
With Tetrad Metals. Two Forms.	$Fe_2(NO_3)_6$	$Fe_2(SO_4)_3$	$Fe_2(PO_4)_2$	

Rules for Writing Chemical Equations.

RULE I.—*As the first member, write the formula of one molecule of each substance taking part in the reaction.*

RULE II.—*As the second member, write the formula of one molecule of each substance observed, or known to be produced during the experiment.*

RULE III.—*Correct the second member, if necessary, by increasing the number of molecules so as to exhaust the supply of elements in the first member.*

RULE IV.—*Correct the first member, if necessary, by increasing the number of molecules absolutely demanded by the substances formed in the second member.*

RULE V.—*Cancel on both sides of the equation— beginning with the first member—all those elements that are used in both members.*

RULE VI.—*See if any elements are left over, after the cancellation required by Rule V. If there are such, combine them in accordance with their known chemical affinities.*

CHAPTER I.

THE NON-METALLIC MONADS.

Hydrogen and Fluorine;
Chlorine, Bromine, and Iodine.

OUTLINE OF THE CHAPTER.

Hydrogen.

    Its distribution in nature, and in the arts.

    Its preparation; by Potassium; by Sodium; by Zinc.

    When it burns it forms Water-vapor, H_2O.

Fluorine.

    Its distribution.—It etches glass.

Chlorine.

    Its distribution in common salt.

    Its preparation from Hydrochloric acid with Manganese di-oxide.

    It is a bleaching agent (because of its affinity for Hydrogen).

    It forms metallic chlorides.

      Hydrochloric Acid.

      Its preparation and properties.

Bromine and Iodine.

    Distribution.—Preparation. – Properties.

THE NON-METALLIC MONADS.

35. Hydrogen is adopted as a monad. In other cases a monad is an element, that—atom for atom—can unite with, or take the place of, Hydrogen.

36. The non-metallic monads are the following:

Name.	Symbol.	Ordinary condition.	Color.	Approximate Atomic weight.
Hydrogen,	H,	gas,	none,	1.
Fluorine,	F,	gas,	none,	19.
Chlorine,	Cl,	gas,	green,	35.4
Bromine,	Br,	liquid,	orange-red,	79.8
Iodine,	I,	solid,	black,	126.6

Hydrogen.

37. The principal *natural* form is in Water, H_2O.

Many *artificial* compounds contain it; thus all acids contain it.

Examples.—Hydrochloric acid,   H Cl.

Sulphuric acid, $H_2 S O_4$.

Nitric acid, $H N O_3$.

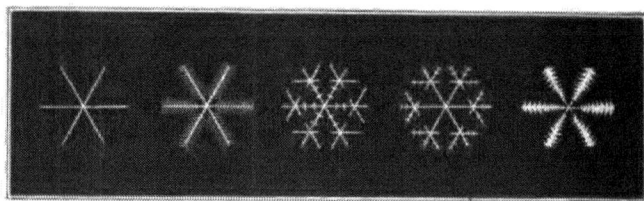

F**IG** 5 —Forms of Water crystallized (as Snow).

38. Potassium liberates Hydrogen. Both take fire.

Experiment.—Place a piece of Potassium upon *dry* filter paper; whittle off the surface and lay the chips aside; throw a small piece of the clean metal upon water in a beaker. Quickly cover the beaker with a piece of glass, or even of paper.

The reaction is,

$$K_2 \; + \; 2 H_2 O \; = \; 2 K O H \; + \; H_2.$$

The **K O H** (Potassium oxy-hydrate, or simply Potassium hydroxide) dissolves in the water; the Hydrogen burns on the surface of the globule of metal; the metal also burns. Thus:

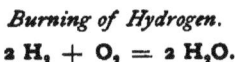

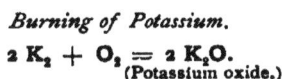

Burning of Hydrogen.

$2 H_2 + O_2 = 2 H_2O.$

Burning of Potassium.

$2 K_2 + O_2 = 2 K_2O.$
(Potassium oxide.)

39. Sodium liberates Hydrogen from cold water; neither of the elements takes fire.

Fig. 6.—Potassium burning, by combining with the Oxygen of water.

Experiment.—Take a piece of Sodium; whittle off the surface and lay the chips aside; throw a fragment on water in a beaker. Quickly cover the beaker with a piece of glass, or even of paper. The metal acts thus:

$$Na_2 + 2 H_2O = 2 Na O H + H_2.$$

(**Na O H** is sodium hydroxide; it dissolves in the water. The Hydrogen escapes, but does not burn.

40. Sodium takes fire on hot water; the liberated Hydrogen also burns.

Experiment.—Try Experiment 39, using hot water; the hot water makes the reaction so violent that sufficient heat is afforded to set on fire both Hydrogen and Sodium. The latter burns with an orange flame.

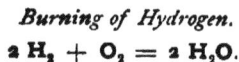

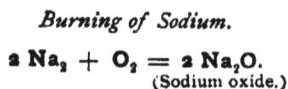

Burning of Hydrogen.

$2 H_2 + O_2 = 2 H_2O.$

Burning of Sodium.

$2 Na_2 + O_2 = 2 Na_2O.$
(Sodium oxide.)

Fig. 7.—Sodium burning on hot water.

41. Sodium, if kept in one place, on cold water, takes fire.

Experiment.—Trim a piece of Sodium as if for Experiment 39. Take a covered beaker of cold water; float a piece of filter-paper on the water; throw a fragment of Sodium upon the wet paper. The wet paper usually keeps the Sodium in one place, so that the heat of the reaction is retained there; the heat thus becomes sufficient to set on fire both Sodium and Hydrogen.

42. Hydrogen, liberated from Water, may be collected.

Experiment.—Fill a large test-tube *full* of water; cover it with a bit of paper; invert it in the water-pan. Trim a piece of Sodium; take

FIG. 8.—Collecting Hydrogen, evolved from Water by Sodium.

it with tweezers; dexterously put it under the mouth of the test-tube. The Sodium will rise in the tube, evolving Hydrogen rapidly. When the reaction ceases, stop the tube with the thumb, hold it with its mouth up, and try the gas with a lighted match. It burns, forming Water vapor, H_2O.

43. Zinc liberates Hydrogen from Sulphuric acid (H_2SO_4).

Experiment.—Fill a small beaker one-fourth full of dilute Sulphuric acid; drop in a few strips of Zinc; cover the beaker with a paper having a half-inch hole in it; hold a lighted match to the Hydrogen gas, escaping at the opening.

$$Zn + H_2SO_4 = H_2 + ZnSO_4 \quad \text{(Zinc Sulphate)}.$$

The Hydrogen unites with Oxygen of the air, and so gives rise to a slight explosion.

$$H_2 + O = H_2O \quad \text{(Water vapor).}$$

Allow the liquid in the beaker to remain on the Zinc for twenty four hours.

44. Zinc and Sulphuric acid form crystals of Zinc sulphate ($ZnSO_4.7 H_2O$).

Experiment.—After the lapse of twenty-four hours—as required by Experiment 43—the solution usually contains a network of crystals of Zinc sulphate. If these crystals fail to appear, repeat Experiment 43, using more Zinc than at the previous trial.

45. Hydrogen liberated from Sulphuric acid (H_2SO_4) may be collected.

Experiment.—Fill a saucer half-full of dilute Sulphuric acid. Also, fill a test-tube *full* of the same, and invert it, while full, into the saucer. Under the mouth of the tube slip a fragment of Zinc and a fragment of Platinum in contact with it. Hydrogen collects in the test-tube. Try it with a lighted taper.

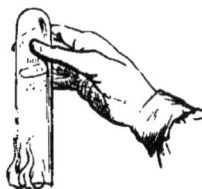

Fluorine, F.

46. Distribution of Fluorine.

The most common *natural* form of Fluorine is the mineral called Fluor-Spar. It is calcium fluoride (CaF_2).

Of the element Fluorine but little is known; it corrodes glass very violently.

FIG. 9.—Hydrogen burning.

The principal *commercial* form of Fluorine is Hydrofluoric acid (HF). It is of itself a gas, but its solution in water is kept in gutta-percha bottles, and is sold in that form.

3*

47. Hydrofluoric acid (HF) attacks glass.

Experiment.—Powder some Fluor-spar; place it in a test-tube; add some concentrated Sulphuric acid, and warm the mixture. Hydrofluoric acid is liberated as a gas.

$$CaF_2 \; + \; H_2SO_4 \; = \; 2\,HF \; + \; CaSO_4 \quad \text{(Calcium sulphate).}$$

The Hydrofluoric acid immediately attacks the glass, corroding and roughening its surface.

48. Hydrofluoric acid may be used for etching glass.

Experiment.—Coat a slip of glass with beeswax over a gentle flame. Scratch some letters through the beeswax to the glass.

Powder some Fluor-spar; place it in a lead saucer; add a considerable quantity of Oil of vitriol (H_2SO_4); place the glass slip on the top of the saucer, and let the whole stand twenty-four hours.

Take off the glass; remove the wax by heating, and then wiping it with a cloth. The inscription should be engraved in the glass by this process.

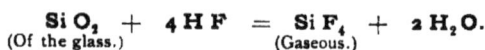

FIG. 10.—Etching glass by means of gaseous Hydrofluoric acid.

$$\underset{\text{(Of the glass.)}}{Si\,O_2} \; + \; 4\,H\,F \; = \; \underset{\text{(Gaseous.)}}{Si\,F_4} \; + \; 2\,H_2O.$$

Chlorine, Cl.

49. Distribution of Chlorine.

In *nature*, Chlorine is never found free; it oftenest occurs in common salt (NaCl, called Sodium chloride). The salt is found in solid deposits, and in the brine of the ocean and of mineral springs.

In *the arts*, Chlorine is largely used in Bleaching-powder, also called Chloride of lime.

50. Preparation of Chlorine.

Experiment.—Prepare Chlorine as follows. Take a deep test tube; place in it some powdered Manganese di-oxide (**Mn O$_2$**, also called Black oxide of manganese). Add some concentrated Hydrochloric acid, and gently warm it for a few minutes. Now place a piece of white paper behind the tube; the greenish color of the gas (and its choking odor) should be detected.

The Chlorine is formed thus:

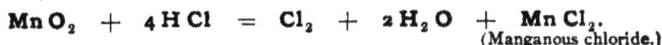

$$Mn\,O_2 \; + \; 4\,H\,Cl \; = \; Cl_2 \; + \; 2\,H_2\,O \; + \; Mn\,Cl_2.$$
(Manganous chloride.)

The gas is more than twice as heavy as air, and it remains in the tube for some time.

51. Chlorine is a bleaching agent, and is used as such, for cotton and linen goods.

Experiment.—Take two small beakers; into one put some dilute Sulphuric acid; into the other put some Bleaching-powder and water.

FIG. 11.—Removing the color from calico by means of Bleaching-powder

Now pass a piece of chocolate calico from one solution to the other, several times; finally wash the cloth in a basin of water. The Sulphuric acid should liberate Chlorine from the Bleaching-powder, and the Chlorine should partly destroy the color.

$$[Ca\,Cl_2\,O_2 \; + \; Ca\,Cl_2] \; + \; 2\,H_2\,SO_4 \; = \; 2\,Cl_2 \; + \; 2\,Ca\,SO_4 \; + \; 2\,H_2O.$$
(Bleaching-powder.)

52. Hydrochloric acid precipitates Silver as Silver chloride (AgCl).

Experiment.—To a solution of Silver nitrate ($AgNO_3$), add a few drops of dilute Hydrochloric acid. A white precipitate of Silver chloride appears.

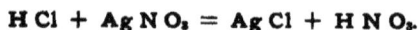

$$H\,Cl + Ag\,N\,O_3 = Ag\,Cl + H\,N\,O_3.$$

53. Common salt precipitates Silver as Silver chloride (AgCl).

Experiment.—To a solution of Silver nitrate, add a dilute solution of common salt (Sodium chloride, $NaCl$). A white precipitate of Silver chloride appears.

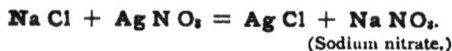

$$Na\,Cl + Ag\,N\,O_3 = Ag\,Cl + Na\,NO_3.$$
$$\text{(Sodium nitrate.)}$$

54. Sunlight decomposes Silver chloride, and blackens it.

Experiment.—Filter the product of the preceding experiment, and expose the white precipitate to the sunlight for twelve hours. The sunlight should decompose it and turn it violet, and finally black.

(It is Silver chloride, on the surface of the paper of a photographic "proof," that becomes black, by exposure to sunlight.)

Hydrochloric Acid, H Cl.

55. Preparation of Hydrochloric acid.

Experiment.—Place a little common salt ($Na\,Cl$) in a small retort; to it, add enough concentrated Sulphuric acid to make a thin paste; con·. nect the neck of the retort with a clean test-tube containing a few drops of water. Then gently heat the retort; Hydrochloric acid ($H\,Cl$) will be formed, and will distill from the retort, and condense in the receiver. Re serve the acid for examination, as described in paragraph 56.

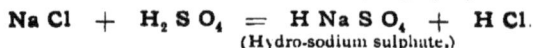

$$Na\,Cl + H_2\,S\,O_4 = H\,Na\,S\,O_4 + H\,Cl.$$
$$\text{(Hydro-sodium sulphate.)}$$

FIG. 12.—Preparation of Hydrochloric acid.

56. Three tests for Hydrochloric acid.

Experiment.—Examine, by three tests, the small amount of Hydro-chloric acid formed:

(*a*) Take a drop on a glass rod, and apply it to blue litmus-paper. It mould turn the paper red.

(*b*) Touch a minute drop to the tongue, and observe the sour taste.

(*c*) Touch a drop to a solution of Silver nitrate, in a test-tube, and observe the white precipitate of Silver chloride formed.

Bromine, Br.

57. Distribution of Bromine.

In *nature*, Bromine is comparatively rare. It is never found free. In sea-water and in saline springs, it occurs as a *Bromide* of certain metals.

In *the arts*, it is known both as Bromine and as Potassium bromide (K Br, also called Bromide of potassium).

58. Preparation of Bromine.

Experiment.—In a deep test-tube, place some Manganese di-oxide and some Potassium bromide. Add a little water to dissolve the latter substance. Next, add some Hydrochloric acid. Now heat the whole, gently. Reddish fumes, and the choking odor of Bromine, should appear.

$$2\,KBr + MnO_2 + 4\,HCl = Br_2 + 2\,H_2O + 2\,KCl + Mn\,Cl_2.$$
(Manganous chloride.)

59. Potassium bromide precipitates Silver as Silver bromide; the product blackens in sunlight.

Experiment.—To a solution of Silver nitrate, add a few drops of solution of Potassium bromide. A yellowish-white precipitate of Silver bromide should appear.

$$K\,Br + Ag\,NO_3 = Ag\,Br + K\,NO_3.$$

Filter, and expose the precipitate to sunlight, for twelve hours. It should blacken, as Silver chloride did. (Experiment 54.)

Iodine, I.

60. Distribution of Iodine.

In *nature*, Iodine is comparatively rare. It is never found free. In sea-water and in saline springs, it occurs as an Iodide of certain metals.

In *the arts*, it is known both as Iodine and as Potassium iodide (KI, also called Iodide of potassium).

61. Preparation of Iodine.

Experiment.—In a deep test-tube, place some Manganese di-oxide and some Potassium iodide. Add a little water to dissolve the latter substance. Now add some Hydrochloric acid, and heat the mixture. A violet vapor of Iodine should arise, and should form—in some part of the tube—a black deposit of solid Iodine.

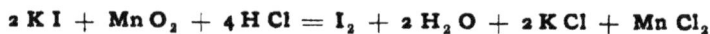

$$2\,KI + MnO_2 + 4\,HCl = I_2 + 2\,H_2O + 2\,KCl + Mn\,Cl_2$$

62. Iodine, when heated, forms a violet vapor which condenses to a black solid.

Experiment.—Heat some fragments of Iodine in a clean test-tube. Observe the heavy violet vapors and the black sublimate or deposit.

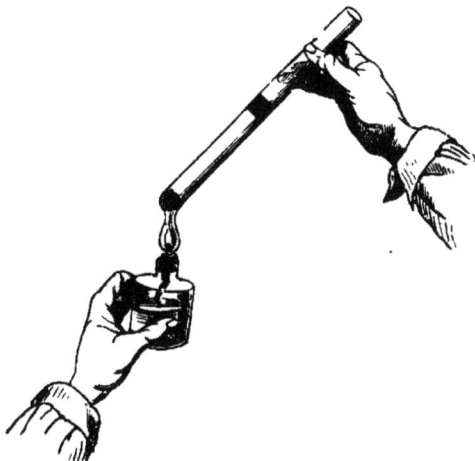

FIG. 13.—Subliming Iodine.

63. Iodine dissolves in alcohol; but it does not dissolve well in water, unless Potassic iodide is present.

Experiment.—Take three clean beakers, and place them upon a white surface.

(*a*) To the first, add a little water.

(*b*) To the second, add alcohol.

(*c*) To the third, add a solution of potassium iodide in water.

To each, add a few fragments of solid Iodine, and observe the different rates of solution of the Iodine.

Save all three for the next Experiment.

64. Under proper conditions, Starch ($C_6H_{10}O_5$) is a very delicate test for *free* Iodine, but not for *combined* Iodine.

Experiment.—Boil a single fragment of Starch, in a tube half-full of water; fill up with cold water; divide this liquid into four parts; to three of them add respectively (*a*), (*b*), and (*c*) of Experiment 63. The difference in the amount of blue color produced, shows a difference in the amount of *free* Iodine dissolved.

Now dissolve a fragment of Potassium iodide in water, and add it to the *fourth* portion of starch-water. It should afford no change of color.

65. Potassium iodide precipitates Silver as Silver iodide; it blackens in sunlight.

Experiment.—To a solution of Silver nitrate, add a few drops of solution of Potassium iodide. A yellowish precipitate of Silver iodide should appear.

$$KI + Ag NO_3 = Ag I + K N O_3$$

FIG. 14.—A precipitate of Silver iodide.

Filter, and expose the precipitate to sunlight for twelve hours. It should blacken, as the Silver chloride and Silver bromide did. (Experiments 54 and 59.)

CHAPTER II.

THE NON-METALLIC DYADS.

Oxygen;
Sulphur, Selenium, and Tellurium.

OUTLINE OF THE CHAPTER.

Oxygen.

    Distribution (most abundant element of our planet);

    Preparation; from Mercuric oxide ($Hg\,O$);

             from $K\,Cl\,O_3$, mixed with $Mn\,O_2$.

It is an energetic supporter of combustion.

Sulphur.

    Distribution; preparation from Pyrites;

    Its properties, shown by heating;

             by dissolving;

             by burning.

Sulphuric Acid.

    Heats water; reddens litmus;

    Precipitates $Pb\,SO_4$, by dilution;

    Chars sugar, starch, paper;

    Dissolves indigo

Sulphuretted Hydrogen.

    Preparation; blackens Lead salts.

Selenium and **Tellurium** are rare.

4

THE NON-METALLIC DYADS.

66. These are the following:

Name.	Symbol.	Ordinary condition.	Color.	Approximate Atomic weight.
Oxygen,	O,	gas,	none,	16.
Sulphur,	S,	solid,	yellow,	32.
Selenium,	Se,	solid,	black,	78.8
Tellurium,	Te,	solid,	white,	128.

Oxygen, O.

67. Oxygen is the most *abundant* element in the earth. It makes up one-half, by weight, of our entire planet. It is also very *widely* distributed.

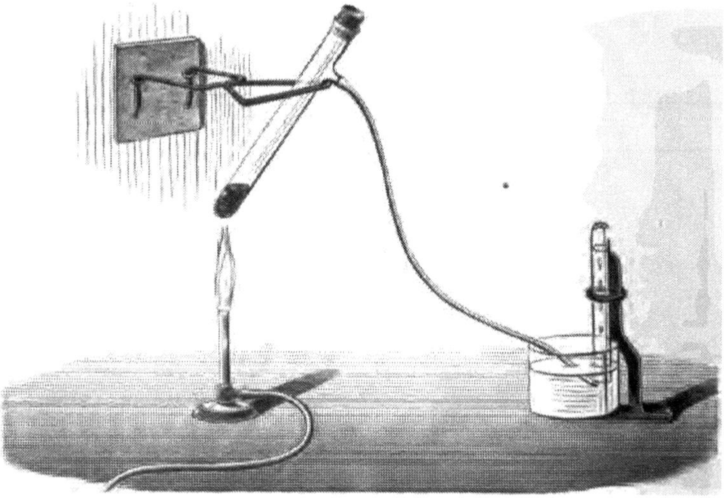

Fig. 15.—Preparation of Oxygen from Mercuric oxide.

It therefore may be said to be found in the majority of substances known.

68. The discoverer of Oxygen, Dr. Priestley, prepared the gas by heating Mercuric oxide (HgO).

Experiment.—Arrange a test-tube as a bell-glass of water, in the water-pan. Put an inch of Red oxide of mercury (Mercuric oxide, **Hg O**) into a fitted 8-inch combustion tube, or one with a side-neck. In either case, the combustion tube must be of very hard glass. Heat the Mercuric oxide carefully, and conduct the Oxygen gas evolved, into the little bell-glass. Try the gas in the bell, by a wax taper which has a spark on it; the gas should relight the taper, and the taper should burn with un usual brilliancy.

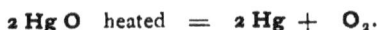

$$2\,Hg\,O \quad \text{heated} \quad = \quad 2\,Hg + O_2.$$

FIG. 16.—Preparation of Oxygen from Mercuric oxide.

69. Oxygen is best prepared from Potassium chlorate, mixed with Manganese di-oxide.

Experiment.—Arrange a test-tube bell in the water-pan. In a small glass retort, place about a teaspoonful of a mixture of about *one part* of Manganese di-oxide, and *three parts* of Potassium chlorate. Now heat

the mixture, and—after some of the atmospheric air has expanded and
passed out of the retort—collect the Oxygen gas in four small bell-glasses.
(A convenient method is to collect the gas in small, wide-mouth bottles.
As each bottle is filled and set aside, cover it with a piece of wet filter-
paper. Reserve the gas for the following experiments: 70, 71, 72, 73.)

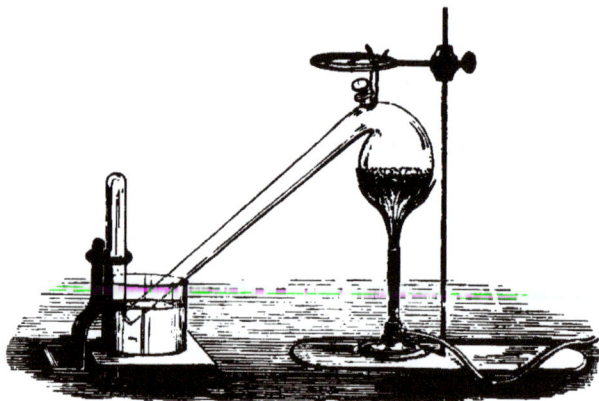

FIG. 17.—Preparation of Oxygen from a mixture of Potassic chlorate and
Manganese di-oxide.

The chemical change may be expressed as follows:

$$2\,K\,Cl\,O_3 \quad \text{heated} \quad = \quad 2\,K\,Cl \quad + \quad 3\,O_2.$$

The Manganese di-oxide undergoes no decided chemical
change in the experiment—indeed, other substances may
be substituted for it. It serves, mainly, to equalize the
application of the heat, and so to prevent the explosive
decomposition of the whole of the Potassium chlorate at
once.

70. Oxygen stimulates the combustion of a candle.

Experiment.—Try one of the jars of Oxygen by a taper having
only a spark upon it; the gas should promptly relight the taper. (See
Experiment 68.)

71. Sulphur burns in Oxygen with a brilliant violet flame. It forms a choking gas, called Sulphurous oxide (SO_2).

Experiment.—Take a fragment of black-board crayon; hollow it, at one end, into a little cup; tie a piece of wire to the cup. In the cup place a fragment of Sulphur. After setting the Sulphur on fire, immerse it in a small jar of Oxygen gas. The Sulphur burns with greatly increased brilliancy.

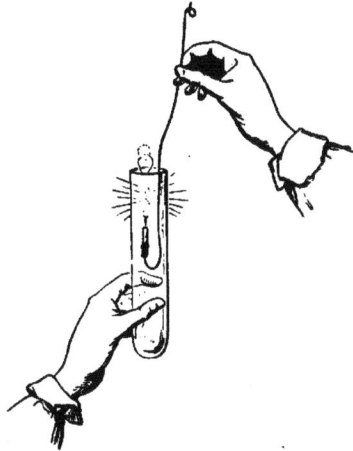

72. Charcoal burns in Oxygen with great brilliancy. It forms Carbon di-oxide (CO_2), a colorless gas.

FIG. 18.—A candle burning in Oxygen.

Experiment.—Twist a bit of wire about a piece of charcoal *bark*. Set one corner of the charcoal on fire by holding it in a lamp-flame. It will not burn freely. Immerse it (when combustion has commenced) in a small bell of Oxygen. The charcoal burns freely and with great brilliancy.

73. Iron burns freely in Oxygen gas. It forms a solid product called **Magnetic oxide**, also called **Ferroso-ferric oxide** (Fe_3O_4).

Experiment.—Twist into a bunch some fine iron wire, called pianoforte wire. (It is the fine wire used by florists.) To one end of the wire attach a fragment of Sulphur. Set the Sulphur on fire, and quickly immerse it in one of the jars of Oxygen. The Sulphur, burning brilliantly, should set the Iron on fire.

74. Nitrates, when heated on charcoal, burn the coal.

Experiment.—Heat, on charcoal, before the blow-pipe, *with care*, a fragment of Potassium nitrate ($K N O_3$). The Oxygen of the Nitrate burns the coal vividly.

4 *

75. Chlorates, when heated on charcoal, burn the coal.

Experiment.—Heat, on charcoal, before the blow-pipe, *with great care*, a fragment of Potassium chlorate ($K\,Cl\,O_3$).

The Oxygen is liberated from the chlorate, and burns the coal with great violence.

Sulphur, S.

76. Distribution of Sulphur.

In *nature*, Sulphur is found free, called *native* Sulphur; it is also found combined with metals, as in Iron pyrites (FeS_2).

In *the arts*, it is known as Flowers of sulphur, and as Roll brimstone; and in many compounds, for example, Sulphuric acid (H_2SO_4).

77. Preparation of Sulphur.

Experiment.—Heat a fragment of Iron pyrites ($Fe\,S_2$) in a blow-pipe tube made of hard glass. The mineral gives off a part of its Sulphur, which collects, as a yellow solid, a little farther up in the tube.

78. Preparation of *brittle* Sulphur.

Experiment.—In a dry test-tube, heat a fragment of Brimstone, until it just fuses. Now pour it into cold water. The cooled Brimstone is brittle.

79. Preparation of *soft* Sulphur.

Experiment.—Heat another portion of Brimstone until it melts; then until it grows thick and dark; then heat further, until it grows thin again; now pour it into cold water. This cooled product is Sulphur, but it is plastic and very different from the product of Experiment 78. (Take care that the Sulphur does not take fire.)

80. Preparation of *crystallized* Sulphur.

Experiment.—Dissolve some Flowers of sulphur in a small quantity of Carbon di-sulphide ($C\,S_2$). Allow the solution to evaporate, by itself, over-night. The Sulphur will be deposited in crystals, from its solution. (Take care that Carbon di-sulphide does not take fire.)

81. Sulphur dioxide (SO_2) is a bleaching agent.

Experiment.—Put a few fragments of Roll brimstone in a small crucible. Heat it carefully until the Sulphur takes fire. Cover the burning Sulphur with a glass lamp-chimney. In the top of the chimney hang a moist carnation-pink or a rosebud. The gas has a slight bleaching action upon the flower.

FIG. 19.—Sulphurous di-oxide bleaching a flower.

The gas is Sulphurous anhydride (SO_2). It is used for bleaching straw and woolen goods.

Sulphuric Acid, $H_2 S O_4$.

82. Sulphuric acid, when mixed with water, produces heat.

Experiment.—Place in a beaker about one fluid-ounce of water; now add, *very carefully*, about four fluid-ounces of concentrated Sulphuric acid. Observe the great heat afforded by the mixture.

83. Sulphuric acid strongly reddens litmus.

Experiment.—Pulverize a few blocks of litmus; add some water to it; add one drop of Sodium hydroxide solution — this gives a blue solu- ion. Now add a drop of Sulphuric acid — this should turn the color red. Now add just enough Sodium hydroxide to turn the color back to blue; finally, add just enough Sulphuric acid to bring the red again.

(Litmus is turned red by acids, and blue by alkalies.)

84. Concentrated Sulphuric acid usually contains Lead sulphate ($PbSO_4$), which it derives from the leaden walls of the large rooms in which it is formed.

Experiment.—Take five fluid-ounces of water. Carefully add one fluid-ounce Sulphuric acid. Stir, and allow to stand over-night. The concentrated acid contains some Lead sulphate ($PbSO_4$) dissolved in it, but this separates from the diluted acid, and is found as a white sediment at the bottom of the beaker.

Save the clear liquid.

85. Concentrated Sulphuric acid chars sugar ($C_{12}H_{22}O_{11}$).

Experiment.—In a beaker, of the size of a tea-cup, place four tea-spoonfuls of white sugar ; add one fluid-ounce of boiling water. Having placed the beaker in a dinner-plate, add, *very carefully*, one ounce of concentrated Sulphuric acid. A black carbonaceous mass appears.

The sugar is composed of Carbon, Hydrogen, and Oxygen. The Sulphuric acid withdraws the Hydrogen and Oxygen—as water—and so leaves the carbonaceous mass.

86. Concentrated Sulphuric acid chars starch ($C_6H_{10}O_5$).

Experiment.—Try the same experiment as 85, only use starch instead of sugar. Starch has the same chemical elements (**C, H,** and **O)** that sugar has. The result is similar.

FIG. 20.—Sulphuric acid charring paper.

87. Dilute Sulphuric acid—when made strong by drying off the water—chars paper ($C_6H_{10}O_5$).

Experiment.—With a quill pen, write, not with ink, but with the acid of Experiment 84, some characters upon white paper. Dry the paper carefully over the lamp-flame. Where the characters are, the paper will become black and charred.

The paper has the same chemical elements (**C**, **H**, and **O**) that starch and sugar have. Here, also, the Sulphuric acid, *when by drying it becomes strong enough*, acts just as in Experiments 85 and 86.

88. Concentrated Sulphuric acid dissolves indigo.

Experiment.—Grind some indigo to a very fine powder. Mix it with clean sand, to prevent the formation of clots of the indigo; add some concentrated Sulphuric acid; allow the whole to stand twenty-four hours; then pour into a half-pint of water.

Filter, and save the blue solution.

Sulphuretted Hydrogen, H_2S.

89. Sulphuretted hydrogen is a colorless gas with a disagreeable odor.

Experiment.—Place in a long test-tube a fragment of Ferrous sulphide (**Fe S**); add a little dilute Sulphuric acid; observe the odor of the gas that is liberated.

$$Fe\,S + H_2SO_4 = H_2S + Fe\,S\,O_4.$$

(Proceed immediately to Experiment 90.)

90. Sulphuretted hydrogen blackens Lead compounds, forming black PbS.

FIG. 21.—Sulphuretted hydrogen blackening Plumbic acetate.

Experiment.—Cover the test-tube (Experiment 89) with a piece of filter-paper which has had a few drops of solution of Lead acetate poured upon it. A black coloration of Lead sulphide (**Pb S**) should appear on the paper.

Selenium, Se, and Tellurium, Te.

91. These elements are so rare that they need not be discussed here.

CHAPTER III.

THE NON-METALLIC TRIADS.

Boron and Nitrogen;
Phosphorus, Arsenic, and Antimony.

.

OUTLINE OF THE CHAPTER.

Boron.

Distribution.

Boric Acid $(H_3 B O_3)$.

Preparation, and flame color.

Nitrogen.

Distribution, in the atmosphere, in Nitrates, etc. Its inertness.

Compounds of Nitrogen and Hydrogen.

Ammonia-gas $(N H_3)$.

When free, gives test with $H Cl$; otherwise, does not.

Ammonium hydroxide $(N H_4 O H)$ is an Alkali.

Ammonium salts are volatile.

Compounds of Nitrogen and Oxygen.

Nitric Acid $(H N O_3)$.

Attacks *quill, indigo, copper, zinc, iron, nickel coin, lead;*

Does not attack *gold;*

Turns Ferrous sulphate, brown.

Preparation, from $K N O_3$ and $H_2 S O_4$. (The product tested.]

Phosphorus.

Occurs in bones; is very combustible;

Burns into $P_2 O_5$; forms Phosphoric acid $(H_3 PO_4)$.

46

Arsenic.

> Occurs in ores; **As$_2$O$_3$** volatilizes readily; is decomposed by carbon;
>
> Forms yellow Arsenious sulphide (**As$_2$S$_3$**).

Antimony.

> Occurs in ores; fuses readily;
>
> Does not dissolve in **H N O$_3$**;
>
> Forms orange Antimonious sulphide (**Sb$_2$S$_3$**).

THE NON-METALLIC TRIADS.

92. The non-metallic triads are the following:

Name.	Symbol.	Ordinary condition.	Color.	Approximate Atomic weight.
Boron,	**B,**	solid,	brown,	10.9
Nitrogen,	**N,**	gas,	none,	14.
Phosphorus,	**P,**	solid,	amber,	31.
Arsenic,	**As,**	solid,	steel,	74.9
Antimony,	**Sb,**	solid,	bluish-white,	120.

Boron, B.

93. Distribution of Boron.

In *nature* and in *the arts*, Boron is little known except in Boric acid (H$_3$BO$_3$) and in Borax (Na$_2$B$_4$O$_7$.10 H$_2$O).

94. Boric acid is a crystalline solid.

Experiment.—Dissolve some Borax in hot water; filter if necessary; add some Hydrochloric acid; allow the whole to cool. White crystals of Boric acid should separate. The Hydrochloric acid sets free 'he weaker Boric acid.

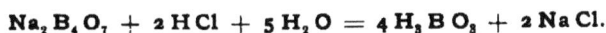

$$Na_2 B_4 O_7 + 2 HCl + 5 H_2 O = 4 H_3 B O_3 + 2 NaCl.$$

95. Boric acid, when highly heated, gives out green light.

Experiment.—Place a little Borax in a casserole, and add some Sulphuric acid to liberate the Boric acid; now add some alcohol; dip a glass rod into the mixture, and then hold the rod in the flame of a lamp. The highly heated Boric acid imparts a delicate green color to the flame. (If the alcohol takes fire in the casserole, and it is desired to extinguish it, cover it with a folded towel.)

FIG. 22. — Boric acid imparts a green color to the flame of alcohol.

Nitrogen, N.

96. Distribution of Nitrogen.

In *nature*, Nitrogen is found in great abundance in our atmosphere. It is also found in Saltpetre (KNO_3, also called Potassium nitrate).

In *the arts*, it exists in Nitrates and Nitric acid (HNO_3).

The *element* Nitrogen is very inert, so that few experiments can be performed with it; but some of its compounds are exceedingly active.

Compounds of Nitrogen and Hydrogen.

$N H_2$, or $N_2 H_4$, called Hydrazine.

$N H_3$, Ammonia-gas.

$N H_4$. Ammonium (existing only in combination with other elements.

Ammonia-gas, NH_3.

97. The test for *free* Ammonia.

Experiment.—Pour some Spirits of Hartshorn (Ammonium hydroxide, NH_4OH) into a small flask; shake the flask; the Ammonium hydroxide gives off colorless, pungent-smelling Ammonia-gas (NH_3). Suspend in the upper part of the flask a glass rod previously dipped in concentrated Hydrochloric acid. Fumes of Ammonium chloride (NH_4Cl) appear.

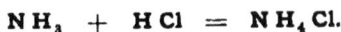

$$NH_3 + HCl = NH_4Cl.$$

The fumes are minute particles of a white solid, NH_4Cl.

Fig. 23.—Ammonia-gas and Hydrochloric-gas meeting in the air and forming Ammonium chloride.

98. Another method of producing the cloud of Ammonium chloride.

Experiment.—Place in a wine-glass or beaker some strong solution of Ammonium hydroxide. Place near it another similar vessel, containing concentrated Hydrochloric acid. A cloud of Ammonium chloride forms in the air between them, especially noticeable when the two glasses are moved from side to side.

99. To test for *combined* Ammonia, having first liberated it.

Experiment.—Into a small flask pour a small quantity of solution of Ammonium chloride (NH_4Cl). Try with the rod and Hydrochloric acid. There should be little, if any, fume. Now add solution of Sodium

5   D

hydrate, and, after shaking the flask, try the rod again. Fumes should now appear, because Ammonia-gas (NH_3) has been liberated, thus :

$$NH_4\,Cl\ +\ NaOH\ =\ NH_3\ +\ H_2O\ +\ NaCl.$$

100. Ammonium hydroxide is an alkali.

Experiment.—Pulverize a block of litmus ; add some water and a drop of dilute Sulphuric acid. The solution will be red ; now, by carefully adding Ammonium hydroxide, the Sulphuric acid may be *neutralized*, and the litmus changed to blue.

101. Ammonia-gas has a very strong attraction for water.

Experiment.—FIRST STAGE.—Invert a small flask in a metal support of some kind, and then fill the flask with Ammonia-gas, by displacement of the air, as follows. In a side-neck test-tube place some strong Aqua-ammonia. Add some Sodium hydroxide (solid or in solution). Now heat the test-tube. By means of a rubber tube, direct the Ammonia-gas upward into the inverted flask.

SECOND STAGE.—When the flask is supposed to be full of Ammonia-gas, place in its neck a perforated cork, fitted with a little glass tube open at both ends. Dip the outer (and larger) opening into a dilute solution of Cupric sulphate. If the experiment is properly conducted, the copper solution will soon be drawn up into the flask, so as to make a miniature fountain. The rapid absorption of the gas by the water-solution causes the latter to be readily forced up by the atmospheric pressure.

102. All Ammonium salts are volatile.

Experiment.—Place a fragment of dry Ammonium chloride on Platinum foil ; heat it over the lamp ; the Ammonium chloride will go off as a vapor, which finally solidifies as dense white smoke.

Compounds of Nitrogen and Oxygen.

N_2O,	Nitrogen protoxide (called laughing-gas).
N_2O_2 (or NO),	Nitrogen di-oxide.
N_2O_3,	Nitrogen tri-oxide or Nitrous anhydride (forms Nitrous acid).
N_2O_4 (or NO_2),	Nitrogen tetroxide (brown fumes).
N_2O_5,	Nitrogen pentoxide or Nitric anhydride (forms Nitric acid).

Nitrogen di-oxide, $N_2 O_2$ (or $N O$).

103. Nitrogen di-oxide, a colorless gas, readily absorbs Oxygen from the air, and then forms brown fumes of Nitrogen tetroxide, N_2O_4 (or NO_2).

Experiment.—Place some copper wire in a side-neck flask. To it, add concentrated Nitric acid. Allow the gas formed to pass into a small bell-glass full of water. (If any brown fumes pass into the bell, they may be disregarded, for they will soon be absorbed by the water.) Finally, empty the little bell-glass into the air; brown fumes will at once appear.

Nitrogen pentoxide, $N_2 O_5$.

104. This substance is often called Nitric anhydride, because it is viewed as Nitric acid deprived of water. With water it forms Nitric acid.

$$N_2 O_5 \ + \ H_2 O \ = \ 2 H N O_3.$$

Nitric Acid, $H N O_3$.

105. Nitric acid turns quill, and other animal matters, to a yellow color.

Experiment.—Warm a few fragments of white quill in dilute Nitric acid; then wash the pieces in water. They will be found to have acquired a permanent yellow color. Many animal matters are turned yellow by Nitric acid.

106. Nitric acid turns indigo to a permanent yellow color.

Experiment.—Place a drop of concentrated Nitric acid on a piece of dark-blue flannel; if the goods are dyed with indigo, the acid produces a bright yellow spot.

107. Nitric acid attacks Copper with violence; it forms a green (or blue) solution; it liberates a gas, at first colorless, then brown.

Experiment.—Place in a test-tube a short strip of Copper wire; add Nitric acid; then warm it gently until the Copper disappears. Cupric nitrate, $Cu(NO_3)_2$, will be formed.

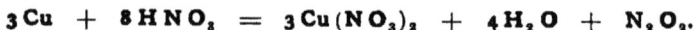

$$3\,Cu\ +\ 8\,HNO_3\ =\ 3\,Cu(NO_3)_2\ +\ 4\,H_2O\ +\ N_2O_2.$$

The action produces a colorless gas (Nitrogen di-oxide, N_2O_2), but this gas, upon coming in contact with the air, combines with Oxygen of the air, and forms brown fumes of Nitrogen tetroxide (N_2O_4), which are seen at the mouth of the tube. (See paragraph 103.)

$$N_2O_2\ +\ O_2\ =\ N_2O_4.$$

108. Nitric acid attacks Zinc with great violence.

Experiment.—Try the same experiment as 107, only employ Zinc in place of Copper. Zinc nitrate, $Zn(NO_3)_2$, will be formed. It gives a colorless solution. It evolves brown fumes.

FIG. 24.—Nitric acid dissolving Copper.

109. Nitric acid attacks Iron with violence.

Experiment.—Try the same experiment as 107, only employ Iron wire in place of Copper. A more complex compound—Ferric nitrate, $Fe_2(NO_3)_6$—is formed.

110. Nitric acid dissolves a Nickel coin.

Experiment.—Try the same experiment as 107, only employ a Nickel coin. As the coin consists of Copper and Nickel or of Copper, Nickel, and Zinc, there may be formed Cupric nitrate, $Cu(NO_3)_2$, Zinc nitrate, $Zn(NO_3)_2$, and Nickelous nitrate, $Ni(NO_3)_2$.

111. Lead and some other metals dissolve better in dilute Nitric acid than in concentrated Nitric acid.

Experiment.—Add concentrated Nitric acid to some shavings of metallic Lead. A *part* of the Lead dissolves, but in so doing it forms crystals of Lead nitrate, $Pb(NO_3)_2$, which collect on the Lead and cover it up, and stop further action of the acid. Now add water; this will be found to dissolve the crystals, and to allow the action of the Nitric acid to continue.

112. Aqua-regia dissolves Gold—though neither of the components, when separate, will do so.

Experiment.—Provide two beakers; into one put some Nitric acid and a strip of Gold-leaf; into the other put some Hydrochloric acid and a strip of Gold-leaf; warm each one separately. The Gold will not dissolve in either case. Mix the contents of the two beakers, and the Gold dissolves at once.

The mixture of these two acids is called *Aqua-regia* (royal-water), because it dissolves the king of the metals, Gold. Auric chloride ($AuCl_3$) is formed.

113. Ferrous sulphate affords a delicate test for Nitric acid.

Experiment.—Dilute some Nitric acid with water to about one-tenth of its strength; into the dilute acid drop a crystal of Ferrous sulphate ($FeSO_4$) and some concentrated Sulphuric acid. The crystal becomes surrounded with a deep-brown coloration (Ferrous nitro-sulphate, $FeSO_4, N_2O_2$). (Ferrous sulphate is called, in commerce, copperas, also green vitriol.)

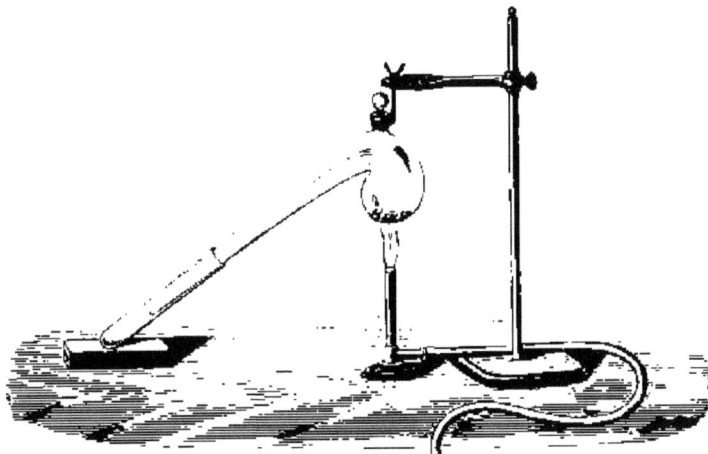

FIG. 25.—Preparation of Nitric acid.

5 *

114. Preparation of Nitric acid (HNO_3).

Experiment.—Place a little Potassium nitrate (KNO_3) in a small retort; to it, add enough concentrated Sulphuric acid to make a thin paste; connect the neck of the retort with a clean test-tube containing a few drops of water, and then gently heat the retort for some time. Nitric acid (HNO_3) will be formed, and will distill from the retort and condense in the receiver.

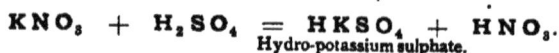

$$KNO_3 + H_2SO_4 = \underset{\text{Hydro-potassium sulphate.}}{HKSO_4} + HNO_3.$$

After a sufficient amount of Nitric acid has collected, test it as described in Experiment 115.

115. Tests for the Nitric acid already prepared.

Experiment.—Divide the acid, produced by Experiment 114, into three parts:

(*a*) To the first portion, add a small fragment of Copper wire; the Copper should freely dissolve, evolving Nitrogen di-oxide, and leaving a blue solution of Cupric nitrate. (See Experiment 107.)

(*b*) To the second portion, add water, and then boil a piece of quill in it. The quill turns yellow. (See Experiment 105.)

(*c*) Dilute the third portion with water, and then add some Sulphuric acid and a crystal of Ferrous sulphate. (See Experiment 113.)

Phosphorus, P.

116. Distribution of Phosphorus.

In *nature*, Phosphorus occurs, principally, in bones. Of dry bone, more than 50 per cent. is Calcium phosphate, $Ca_3(PO_4)_2$.

In *the arts*, the element Phosphorus and many Phosphates are employed.

Phosphorus is very poisonous and very combustible. It should *never be touched* with the hands, since dangerous burns are often caused by it.

117. Phosphorus burns into Phosphoric oxide (P_2O_5).

Experiment.—Cut a piece of Phosphorus under water. Dry, upon a piece of filter-paper, a fragment smaller than a pea; place the fragment upon a piece of wood, metal, or porcelain. After setting the Phosphorus on fire, cover it quickly with a large jar or bell-glass. The white fumes are Phosphoric anhydride (P_2O_5).

118. Phosphoric anhydride, with water, produces Phosphoric acid (H_3PO_4)

Experiment.—The Phosphoric anhydride of the last experiment is a white, snow-like substance which quickly absorbs moisture from the atmosphere—so quickly, in fact, that the white substance cannot always be secured.

FIG. 26.—Phosphorus burning in air.

If a little of it can be obtained, throw it on water; it hisses, owing to the heat of combination. It forms Phosphoric acid (H_3PO_4).

$$3\,H_2O \;+\; P_2O_5 \;=\; 2\,(H_3PO_4).$$

Apply a fragment of moist blue litmus-paper to the place, under the jar, where the Phosphoric acid is supposed to be. A reddening of the paper will indicate the presence of the acid.

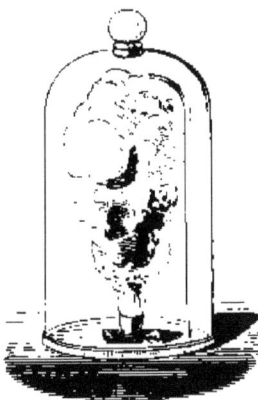

Arsenic, As.

119. Distribution of Arsenic.

In *nature*, Arsenic occurs as a constituent of a large number of metallic ores.

In *the arts*, it occurs in the form of Arsenious oxide (As_2O_3), also called White-arsenic. It is also contained in Paris-green.

Most of the compounds of Arsenic are very poisonous; they must be handled with great care.

120. Arsenical compounds, heated on charcoal, give a peculiar odor.

Experiment.—Heat upon charcoal a fragment of As_2O_3 no bigger than a pin's head. It volatilizes, giving forth an odor of garlic.

121. White-arsenic may be sublimed in a glass tube.

Experiment.—Heat, in a blow-pipe tube of hard glass, a pin-head of White-arsenic. It volatilizes, and condenses in crystals of the same composition as before.

FIG. 27.—Arsenious oxide detected by heated charcoal.

122. Highly heated Carbon withdraws Oxygen from Arsenious oxide.

Experiment.—Place a pin-head of Arsenious oxide in a blow-pipe tube; above it, place a minute fragment of charcoal; heat the tube near the charcoal, and then near the White-arsenic. The Carbon should take

Oxygen from the Arsenious anhydride, forming Carbonic anhydride gas, which escapes, while uncombined Arsenic forms a black sublimate in the tube.

123. Arsenic can form a yellow sulphide of Arsenic (As_2S_3).

Experiment.—Dissolve a pin-head of White-arsenic in Hydrochloric acid; this produces Arsenious chloride ($As\ Cl_3$). Now add water and then Sulphuretted-hydrogen as gas, or dissolved in water. A yellow precipitate of Arsenious sulphide appears ($As_2\ S_3$).

$$2\ As\ Cl_3\ +\ 3\ H_2\ S\ =\ As_2\ S_3\ +\ 6\ H\ Cl.$$

Antimony, Sb. (Stibium.)

124. Distribution of Antimony.

In *nature*, Antimony is found in many minerals, principally in Stibnite (Sb_2S_3).

In *the arts*, it is known as the element Antimony, called *metallic* Antimony. *Tartar emetic* is a double tartrate of Antimony and Potassium.

125. Antimony fuses readily and burns readily.

Experiment.—Fuse, on charcoal, a small fragment of Antimony (not larger than a pin-head). It fuses readily, and, if it drops on the table, the molten fragments hop along, burning in the air, and leaving a small, smoky ash of Antimonic oxide ($Sb_2\ O_5$).

126. Antimony does not dissolve in Nitric acid.

Experiment.—Boil a few fragments of Antimony with a little Nitric acid. The acid does not dissolve the Antimony, though it changes it into an oxide (Antimony tetroxide, $Sb_2\ O_4$).

127. Antimony forms an orange Sulphide of Antimony (Sb_2S_3).

Experiment.—Dissolve a little Tartar emetic in water and a few drops of Hydrochloric acid. Add some Sulphuretted-hydrogen as gas, or dissolved in water. An orange precipitate of Antimonious sulphide appears ($Sb_2 S_3$).

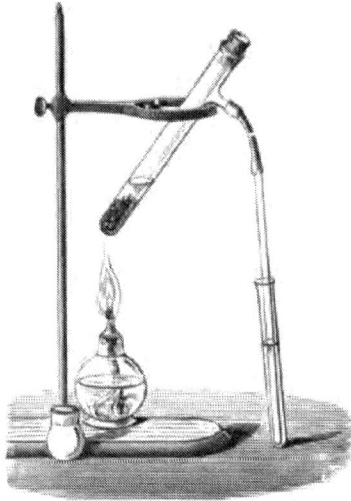

FIG 28.—Sulphuretted-hydrogen gas used to precipitate Antimony.

CHAPTER IV.

•

THE NON-METALLIC TETRADS.

Carbon;
Silicon, Titanium, and Tin.

OUTLINE OF THE CHAPTER.

Carbon.

Distribution in nature;

Shown to exist in starch;

May be detected by Potassium nitrate ($K\,N\,O_b$).

Charcoal decolorizes indigo solution;

Does not dissolve in any ordinary solvent.

Compounds of Carbon and Hydrogen.

Compounds of Carbon, Hydrogen, and Oxygen.

Compounds of Carbon and Oxygen.

Carbon di-oxide ($C\,O_2$).

Is prepared from marble;

Extinguishes flame;

May be poured into another vessel;

Makes lime-water milky;

Is exhaled from the lungs.

Silicon.

Distribution in nature;

Sand ($Si\,O_2$) is difficult to fuse; and to dissolve.

Soluble glass is decomposed by $H\,Cl$.

Titanium.

(Rare.)

Tin.

Distribution.

It dissolves in $H\,Cl$, but not in $H\,N\,O_3$.

THE NON-METALLIC TETRADS.

128. The non-metallic tetrads are the following:

Name.	Symbol.	Ordinary condition.	Color.	Approximate Atomic weight.
Carbon,	**C,**	solid,	black,	12.
Silicon,	**Si,** .	solid,	black,	28.2
Titanium,	**Ti,**	solid,	dark green,	48.
Tin,	**Sn,**	solid,	white,	117.7

Carbon, C.

129. Distribution of Carbon.

In *nature*, Carbon exists (*a*) crystallized in the Dia-mond; (*b*) as Graphite, the black mineral called also Plumbago and Black-lead, and used in lead-pencils; (*c*) as Charcoal, which is formed by heating either animal or vegetable matters in such a way as to expel elements other than Carbon, and to leave the latter.

Pit-coal (Anthracite or Bituminous) is far from pure carbon; it contains many other elements.

130. Charring affords a simple test for Carbon.

Experiment.—Heat a small fragment of starch in a test-tube; an impure carbon is left. (Starch is $C_6H_{10}O_5$.)

131. Potassium nitrate deflagrates with charcoal.

FIG. 29.—Carbon. Side-view and top-view of a diamond, cut in the form called a brilliant.

Experiment.—Fuse gently on platinum a small piece of Potassium nitrate. Carefully drop into the fused mass a fragment of charcoal. Heat

carefully, if necessary. The deflagration which ensues is a true combustion of the coal, the **K N O₃** furnishing the Oxygen.

132. Carbon is a decolorizing agent.

Experiment.—Filter an indigo solution (Experiment 88) through paper. It passes through still blue, showing that we have a true *solution* of the indigo.

To the filtrate, add some animal charcoal, and filter again. The charcoal manifests a wonderful decolorizing power.

133. Carbon is not soluble in any ordinary solvent.

Experiment.—Boil a fragment of charcoal with Hydrochloric acid. It will not dissolve.

There is scarcely any substance known that will dissolve Carbon *as an element* and without changing it into some new compound.

FIG. 30.—Indigo solution decolorized by filtering through animal charcoal.

Compounds of Carbon and Hydrogen.

134. These compounds are numbered by hundreds:

Of *solids*, Paraffin;

Of *liquids*, Turpentine;

Of *gases*, Illuminating gases, are examples.

Test each of these for Carbon, by burning them carefully in or under a porcelain dish, so as to give a deposit of lamp-black.

Two of the best-known gaseous hydro-carbons are Marsh gas, also called Methyl hydride (CH_4), and Olefiant gas, also called Ethylene (C_2H_4).

6

Olefiant gas (*Ethylene*), C_2H_4.

135. Olefiant gas burns with a luminous flame. (It must be prepared *with great care*, one reason being because it forms an explosive mixture with air.)

Experiment.—Place about half a thimbleful of ordinary Alcohol (Ethyl alcohol) in a side-neck flask. To it, add about four times its bulk of concentrated Sulphuric acid; add also a little clean sand, to prevent frothing. Heat the flask, *carefully;* and when the gas appears to have expelled the air of the apparatus, collect what next comes, in a small bell-glass. Afterward try the gas with a lighted taper. It should burn with a yellow flame. (It is not pure Ethylene.)

Compounds of Carbon, Hydrogen, and Oxygen.

136. These, also, are extremely numerous. Starch $(C_6H_{10}O_5)$, Wood $(C_6H_{10}O_5)$, Sugar $(C_{12}H_{22}O_{11})$, and Alcohol (C_2H_5OH), are examples.

Compounds of Carbon and Oxygen.

137. Carbon forms two compounds with Oxygen, namely, Carbon mon-oxide (CO), and Carbon di-oxide (CO_2). Both of them are colorless gases.

Carbon mon-oxide, C O.

138. Carbon mon-oxide burns with a blue flame, and with the production of Carbon di-oxide (CO_2).

Experiment.—Place a few fragments of crystallized Oxalic acid $(H_2O_2C_2O_2)$ in a side-neck tube. To it, add sufficient Sulphuric acid to moisten it. Now heat gently. Carry the evolved gas to a small bell-glass. Afterward try the gas with a lighted taper; it should burn with a pale-blue flame. (*The gas is poisonous.*)

Carbon di-oxide, C O₂.

139. Carbon di-oxide (CO_2) (also called Carbonic acid, Carbonic anhydride, etc.) may be prepared from a Carbonate.

Experiment.—Fill a test-tube one-third full of Hydrochloric acid; drop into the acid a fragment of Potassium carbonate; the effervescence observed is due to the escape of Carbonic anhydride, a gas.

$$K_2 C O_3 + 2 H Cl = C O_2 + 2 K Cl + H_2 O.$$

(White marble, which is Calcium carbonate ($Ca C O_3$), may be used in place of Potassium carbonate. See next experiment, No. 140.)

140. Carbon di-oxide extinguishes flame.

Experiment.—Put a little Hydrochloric acid in the bottom of a wide-mouthed candy-jar or other jar; add some fragments of marble; allow the action to go on for a few minutes. Immerse a candle or a lighted taper in the jar; when it comes below the surface of the Carbonic gas, it will be extinguished suddenly.

$$Ca C O_3 + 2 H Cl = C O_2 + Ca Cl_2 + H_2 O.$$

141. Carbon di-oxide, though invisible, is a heavy gas, and may be poured from one vessel to another.

Experiment.—Suspend a lighted candle in a small beaker or jar, or cover the lighted candle with a glass lamp-chimney. Upon the candle, carefully pour the gas left in the jar from Experiment 140. If the experiment is properly performed, the candle will be quickly extinguished by the falling gas.

Fig. 31.—Pouring Carbon di-oxide from one vessel to another to extinguish a burning candle.

142. Carbon di-oxide makes lime-water milky with Calcium carbonate.

Experiment.—Prepare some fresh lime-water as follows. Pulverize a fragment of quicklime (**Ca O,** called Calcium oxide); then place it in a pint bottle of water. Allow the mixture to stand over-night or until the solid subsides, and the liquid becomes quite clear.

Generate a little Carbonic gas in a beaker, and pour the gas into a smaller beaker half-full of fresh lime-water; on the surface of the lime-water a white precipitate of Calcium carbonate (**Ca C O₃**) appears. Stirring the solution favors absorption of the gas.

$$CaO_2H_2 \; + \; CO_2$$
$$= Ca\,CO_3 \; + \; H_2O.$$

FIG. 32.—Carbonic di-oxide from the lungs passed into Lime-water (**Ca O₂ H₂**).

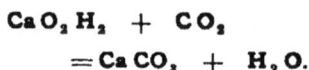

143. Carbon di-oxide is exhaled from the lungs of living animals.

Experiment.—Take a beaker one-third full of fresh and clear lime-water; by means of a glass tube, blow a few bubbles of breath into the lime-water; the Carbonic gas exhaled from the lungs will soon render the clear water milky, with Calcium carbonate.

Silicon, Si.

144. Distribution of Silicon.

In *nature*, Silicon is the second element in order of abundance. One-fourth, by weight, of our planet, is Silicon. But it is extremely difficult to obtain uncombined Silicon, owing to its intense affinity for Oxygen, with which it is almost always united. Sand, quartz, and rock-crystal are forms of Silicic oxide (SiO_2). Most other rocks are Silicates.

In *the arts*, glass and Silicate of soda are its best-known compounds.

145. Sand is very infusible.

Experiment.—Heat some sand on a platinum foil; it will not melt.

FIG. 33.—Crystals of Quartz, Silicic oxide ($Si O_2$).

146. Sand is very insoluble.

Experiment.—Boil some sand in a test-tube with Hydrochloric acid; it scarcely dissolves at all.

147. Silicic acid is gelatinous or jelly-like.

Experiment.—Place some *soluble Silicate of soda* in a test-tube, and then add some Hydrochloric acid; a gelatinous precipitate of Silicic acid is formed.

Titanium, Ti.

(This substance is comparatively rare, and need not be described here.)

6 *   E

Tin, Sn. (Stannum.)

148. Distribution of Tin.

In *nature*, Tin occurs as an oxide, called Stannic oxide (SnO_2).

In *the arts*, the uncombined element is called Block-tin. What is called Sheet-tin is really Sheet-iron with a thin coating of Tin. Stannous chloride ($SnCl_2$), also called Tin-crystals, and Sodium stannate (Na_2SnO_3), are much used in dyeing.

FIG. 34. — Passing Sulphuretted-hydrogen gas into a solution of Tin.

149. Tin dissolves in Hydrochloric acid.

Experiment.—Boil Tin-foil or some fillings of Tin, in Hydrochloric acid; they partly or wholly dissolve, forming **Sn Cl$_2$**.

$$\text{Sn} + 2\,\text{H Cl} = \text{Sn Cl}_2 + \text{H}_2.$$

Test a little of the clear solution by adding Sulphuretted-hydrogen as gas or dissolved in water. The dark-brown precipitate is Stannous sulphide (**Sn S**).

150. Stannous sulphide is a dark-brown precipitate.

Experiment.—Dissolve some Stannous chloride in water and Hydrochloric acid; add some Sulphuretted-hydrogen as gas or dissolved in water; a *dark-brown* precipitate of Stannous sulphide (**Sn S**) appears.

$$\text{Sn Cl}_2 + \text{H}_2\text{S} = \text{Sn S} + 2\,\text{H Cl}.$$

151. Nitric acid oxidizes Tin, but does not dissolve it.

Experiment.—Boil some Tin-foil or filings of Tin, in Nitric acid; the acid does not dissolve the metal, though it changes it to a white insoluble acid, called Meta-stannic acid ($\text{H}_{10}\text{Sn}_5\text{O}_{15}$).

CHAPTER V.

THE METALLIC MONADS.

Silver;
Potassium, Sodium, and Lithium.

OUTLINE OF THE CHAPTER.

Silver.

> Distribution.
> It dissolves in dilute Nitric acid.
> Silver coin proved to contain Copper and Silver.

Potassium.

> Distribution.   Many important Potassic salts are used in the arts.
> Potassium carbonate deliquesces.
> Potassium chlorate deflagrates.
> Potassium nitrate gives a good flame-color.
> Potassium di-chromate forms chrome-yellow.

Sodium.

> Distribution.   A few Sodium salts are used in enormous quantities
> in the arts.
> Sodium salts give an orange flame-color.

Lithium.

> Its salts give a crimson flame-color.

67

THE METALLIC MONADS.

152. The principal metallic monads are the following:

Name.	Symbol.	Ordinary condition.	Color.	Approximate Atomic weight.
Silver,	**Ag,**	solid,	white,	107.7
Potassium,	**K,**	solid,	white,	39.
Sodium,	**Na,**	solid,	white,	23.
Lithium,	**Li,**	solid,	white,	7.

Silver, Ag. (Argentum.)

153. Distribution of Silver.

In *nature*, Silver is found native or uncombined; it also occurs in a state of combination with Sulphur and with other elements.

In *the arts*, Silver coins and Silver ware are employed. They are usually alloys of Silver and Copper, the Copper giving hardness to the alloys. Silver nitrate ($AgNO_3$) —also called Nitrate of silver—is largely used by photographers.

154. Silver dissolves best in *dilute* Nitric acid.

Experiment.—Dissolve a fragment of a silver five-cent piece, by boiling, in dilute Nitric acid; divide the solution into two parts for the next two experiments.

155. First method of testing for Silver and Copper.

Experiment.—To the *first part* of the solution of Experiment 154, add some Hydrochloric acid. This precipitates the Silver as Silver chloride (**Ag Cl**). Filter,

FIG. 35.—Silver coin dissolving in Nitric acid.

and to the filtrate, add Ammonium hydroxide. This forms a deep-blue compound with the Copper, and so shows the presence of the latter metal.

156. Second method of testing for Silver and Copper.

Experiment.—In the *second part* of the solution of Experiment 154, use a solution of common salt, in place of Hydrochloric acid, for precipitating the Silver; continue the experiment as in Experiment 155. Common salt answers the same purpose as Hydrochloric acid, and is cheaper.

157. The metallic Silver may be recovered.

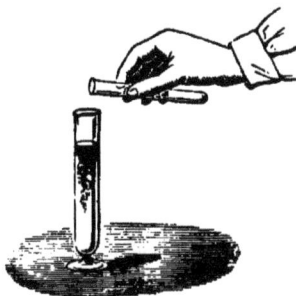

FIG. 36.—Making a precipitate of Argentic chloride.

Experiment.—Remove from the filters of Experiments 155 and 156 the Silver chloride obtained; place it on charcoal, with some dry Potassium carbonate; fuse the mixture with a blow-pipe, until globules of pure Silver are obtained. The Potassium of the Potassium carbonate withdraws Chlorine to form Potassium chloride; the Silver is thus liberated.

Potassium K. (Kalium.)

158. In *nature*, Potassium exists in many minerals. The metal itself is very difficult of preparation because of its intense affinity for Oxygen; even when once prepared, it quickly absorbs Oxygen from air, or even from water. The metal must be preserved under some oil that contains no Oxygen.

In *the arts*, a familiar source of Potassium is wood-ashes. The following-named compounds are well known and largely used—

Potassium carbonate,	$K_2 CO_3$;	
Potassium chlorate,	$K Cl O_3$;	
Potassium di-chromate,	$K_2 Cr_2 O_7$	(also called Bi-chrome);
Potassium hydroxide,	$K O H$	(also called Caustic potash);
Potassium nitrate,	$K N O_3$	(also called Nitre and Saltpetre).

159. Potassium carbonate deliquesces and effervesces.

Experiments.—(*a*) Place a little of the dry Potassium carbonate in a watch-glass, and allow it to stand for twenty-four hours exposed to the open air; it has so strong an attraction for the moisture of the air that it frequently entirely liquefies.

(*b*) Add a little Hydrochloric acid to a few fragments of Potassium carbonate, in a test-tube, and observe the effervescence.

$$K_2CO_3 + 2HCl = CO_2 + 2KCl + H_2O.$$

160. Potassium chlorate ($KClO_3$) deflagrates on charcoal.

Experiments.—(*a*) Fuse a fragment of the dry salt on charcoal; observe the rapid combustion of the coal, due to the Oxygen of the salt.

(*b*) Gently fuse a portion of the salt on clean porcelain; a reaction occurs, but it is hardly perceptible. (See page 40.)

161. Potassium nitrate (KNO_3) gives the violet flame-color of Potassium.

Experiments.—(*a*) and (*b*). Try with this salt two experiments similar to those of Experiment 160.

(*c*) Wet a Platinum wire loop; dip it in powdered Potassium nitrate; then fuse the salt in the Bunsen lamp-flame. Observe the violet Potassium-color.

162. Potassium di-chromate produces chrome-yellow.

Experiment.—Dissolve a fragment of Potassium di-chromate in water, and add the liquid to a solution of Lead acetate. A yellow precipitate of Lead chromate, also called Chrome-yellow ($Pb Cr O_4$) appears. (See Experiment 173.)

Fig. 37.—Producing the violet flame-color of Potassium.

Sodium, Na. (Natrium.)

163. Distribution of Sodium.

In *nature*, Sodium exists in many minerals. The best example is Rock-salt (NaCl).

In *the arts*, metallic Sodium is somewhat used.

Sodium hydroxide (NaOH), called Caustic-soda, is used in the manufacture of soap; *Sodium chloride* (NaCl), common salt, is used for culinary and for manufacturing purposes; *Sodium carbonate* (Na_2CO_3), called Soda-ash, is used in the bleaching of cotton goods, the scouring of wool, and the manufacture of soap and of glass. The consumption of Soda-ash is enormous.

164. Sodium salts afford a peculiar orange flame-color.

Experiment.—Heat, in the lamp-flame, a Platinum wire which has been dipped into some powdered Sodium chloride. Observe the yellow Sodium light; meanwhile, hold near the flame a small bright-red object—*e. g.*, a clear crystal of Potassium di-chromate ($K_2 Cr_2 O_7$), or a small quantity of a very concentrated red solution of the same salt in a test-tube. Notice that the Sodium flame peculiarly degrades the color of the object.

FIG. 38.—Producing the orange flame-color of Sodium.

Lithium, Li.

165. Distribution of Lithium.

Lithium is rare both in *nature* and in *the arts*.

166. Lithium salts afford a crimson flame-color.

Experiment.—Add a drop of Hydrochloric acid to a minute portion of Lithium carbonate, in a watch-glass. Dip a Platinum wire into the solution, and then heat it in the lamp-flame. A magnificent crimson flame is characteristic of Lithium.

CHAPTER VI.

THE METALLIC DYADS.—FIRST SECTION.

Lead;
Barium, Strontium, and Calcium.

OUTLINE OF THE FIRST SECTION.

Lead.

Distribution.

Properties of its salts: Lead chloride, white; Lead iodide, yellow; Lead sulphide, black; Lead chromate, yellow.

Precipitation of metallic Lead, by Zinc.

Barium.

Forms Barium sulphate, which is very insoluble.

Its salts afford green flame-colors.

Strontium.

Forms Strontium sulphate.

Its salts afford red flame-colors.

Calcium.

Distribution.

Properties of Quicklime; of Calcium chloride; of calcium sulphate.

72

THE METALLIC DYADS.—FIRST SECTION.

167. The First Section of the metallic dyads includes the following :

Name.	Symbol.	Ordinary condition.	Color.	Approximate Atomic weight.
Lead,	**Pb,**	solid,	bluish-white,	206.5
Barium,	**Ba,**	solid,	yellow,	136.8
Strontium,	**Sr,**	solid,	yellow,	87.4
Calcium,	**Ca,**	solid,	yellow,	40.

Lead, Pb. (Plumbum.)

168. Distribution of Lead.

Lead occurs, *in nature*, as Galena (Lead sulphide, PbS), and also in many other ores.

In *the arts*, one of its most important uses is for Lead pipe ; another very important use is in the manufacture of White lead (a hydrated Carbonate of lead), which is the basis of nearly all paints.

169. Lead nitrate dissolves in water.

Experiment.—Dissolve some Lead nitrate, **Pb (N O₃)₂**, in water, and divide the solution, so formed, into four parts, for the following four experiments.

170. Lead chloride is insoluble in cold water, but dissolves in hot water.

Experiment.—To solution of Lead nitrate, add Hydrochloric acid ; a white crystalline precipitate of Lead chloride (**PbCl₂**) appears.

$$Pb\,(N\,O_3)_2 \;+\; 2\,H\,Cl \;=\; Pb\,Cl_2 \;+\; 2\,H\,N\,O_3.$$

Allow the precipitate a few moments to subside ; then decant the clear liquid. To the precipitate, add some clean water, and boil ; the precipitate dissolves wholly or in part ; now allow the whole to cool, when the Lead chloride that dissolved will re-appear as feathery crystals.

7

171. Lead iodide is insoluble in cold water, but dissolves in hot water.

Experiment.—To solution of Lead nitrate, add solution of Potassium iodide ; a yellow precipitate of Lead iodide (Pb I₂) appears.

$$Pb(NO_3)_2 + 2KI = PbI_2 + 2KNO_3.$$

Allow the precipitate a few moments to subside ; then decant the clear liquid. To the precipitate, add some clean water, and boil ; the precipitate dissolves wholly or in part ; now allow the whole to cool, when the Lead iodide will re-appear as golden crystalline spangles.

172. Sulphuretted-hydrogen affords a delicate test for Lead.

Experiment.—To solution of Lead nitrate, add some Sulphuretted-hydrogen as gas, or dissolved in water : a black or brownish-black precipitate of Lead sulphide (PbS) appears.

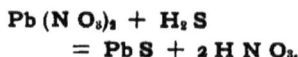

$$Pb(NO_3)_2 + H_2S = PbS + 2HNO_3.$$

173. Potassium di-chromate is used as a test for Lead.

Experiment.—To solution of Lead nitrate, add solution of Potassium di-chromate : a yellow precipitate of Lead chromate (Pb Cr O₄) appears. Allow the precipitate a few moments to subside, and then pour off the clear liquid.

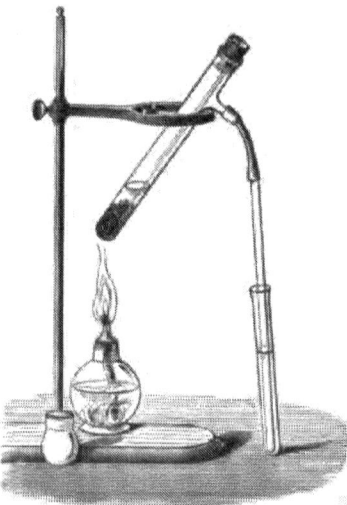

To the precipitate, add solution of Sodium hydroxide until it dissolves ; next add Acetic acid ; this will neutralize the Sodium hydroxide which dissolved the chrome-yellow. The latter will then re-appear.

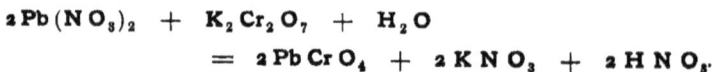

FIG. 39.—Passing Sulphuretted-hydrogen gas into a solution containing Lead.

$$2Pb(NO_3)_2 + K_2Cr_2O_7 + H_2O = 2PbCrO_4 + 2KNO_3 + 2HNO_3.$$

174. Metallic Lead may be liberated from its solutions by metallic Zinc.

Experiment.—Fill a beaker or bottle nearly full of a dilute solution of Lead acetate; in the solution suspend a strip of metallic Zinc. A portion of the Lead is precipitated from the solution in the form of bright metallic flakes upon the Zinc. But, at the same time, there is dissolved an amount of metallic Zinc, that is chemically equivalent to the Lead precipitated.

FIG. 40.—The Lead-tree precipitated by metallic Zinc.

Barium, Ba.

175. Distribution of Barium.

The most abundant *natural* form of Barium is the mineral Heavy-spar. It is Barium sulphate ($BaSO_4$).

In *the arts,* Barium chloride ($BaCl_2$) and Barium nitrate, $Ba(NO_3)_2$ are considerably used.

176. Barium and Sulphuric acid are tests for each other.

Experiment.—Add a drop of dilute Sulphuric acid to a solution of Barium chloride. It gives a milk-white precipitate of Barium sulphate ($BaSO_4$), which is one of the most insoluble of known substances. Hence, Sulphuric acid is used as a test for Barium compounds, and, *vice versâ*, Barium compounds are used as a test for Sulphuric acid.

$$BaCl_2 + H_2SO_4 = BaSO_4 + 2HCl.$$

177. Barium salts afford a green flame-color.

Experiment.—Moisten a Platinum wire loop; dip it in powdered Barium chloride, and then place it in the lamp-flame, and keep it there for some time. Barium salts impart a yellowish-green color to the flame.

178. Barium salts are used to give the color in green-fire.

Experiment.—Pulverize *separately, with great care,*

> Barium nitrate,
> Potassium chlorate,
> Gum shellac.

Then measure, in a small dry test-tube, an equal quantity, by bulk, of each of the three substances. (It will be found convenient to measure the powdered Shellac between the two white powders. The quantities used are thus easier distinguished.)

Now mix the powders gently and carefully, but thoroughly, on a piece of paper. Place the mixture in an iron pan, or on a wooden block, and set fire to it. It affords green-fire.

The Shellac is a vegetable substance and contains Carbon. The combustion of this Carbon is sustained by the Oxygen of the Nitrate and Chlorate. (See Experiments 74 and 75.) At the same time, the Barium imparts the green color to the flame.

FIG. 41.—Green-fire colored by a salt of Barium.

Strontium, Sr.

179. Distribution of Strontium.

In *nature*, Strontium is not very abundant; in *the arts*, Strontium nitrate, $Sr(NO_3)_2$, and Strontium chloride ($SrCl_2$) are somewhat used.

180. Strontium forms a Sulphate resembling Barium sulphate.

Experiment.—To a solution of Strontium nitrate, add some dilute Sulphuric acid; a white precipitate of Strontium sulphate appears.

$$Sr(NO_3)_2 + H_2SO_4 = SrSO_4 + 2HNO_3.$$

181. Strontium salts afford a red flame-color.

Experiment.—Moisten a platinum wire loop; dip it in powdered Strontium nitrate, and then place it in the lamp-flame.

Strontium salts impart a deep-red color to the flame.

182. Strontium salts are used to give the color in red-fire.

Experiment. — Pulverize *separately, with great care,*

Strontium nitrate,
Potassium chlorate,
Gum shellac.

Then measure, in a small dry test-tube, an equal quantity, by bulk, of the three substances. Now mix the powders gently and carefully, but thoroughly, on a piece of paper. Place the mixture in an iron pan, or on a wooden block, and set fire to it. It affords red-fire. (See Experiment 178.)

FIG. 42.—Red-fire colored by a salt of Strontium.

Calcium, Ca.

183. Distribution of Calcium.

Calcium is a very abundant and widely diffused substance.

In *nature*, it is a characteristic constituent of shells, marble, and limestones, also of gypsum, bones, and many other substances.

In *the arts*, it is used in enormous quantities in such compounds as lime, bleaching-powder, etc.

184. Calcium chloride is deliquescent.

Experiment.—Place about a teaspoonful of concentrated Hydrochloric acid in a casserole; drop a piece of litmus-paper into it. Now stir in slaked or unslaked quicklime, little by little, until the acid is entirely neutralized; this point is attained when the litmus-paper becomes blue. Filter the whole mass. The clear filtrate contains the Calcium chloride.

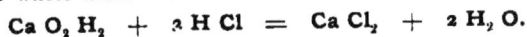

$$Ca\,O_2\,H_2 \; + \; 2\,H\,Cl \; = \; Ca\,Cl_2 \; + \; 2\,H_2\,O.$$

Now evaporate the solution to dryness, and allow the dry residue to remain, for twenty-four hours, exposed to the air. Calcium chloride has so strong an attraction for moisture that it soon absorbs from the atmosphere water enough to liquefy itself.

185. Calcium chloride may be changed back to Calcium carbonate.

Experiment.—Add some water to the Calcium chloride afforded by Experiment 184. Now add Ammonium hydroxide and Ammonium carbonate solution. A white precipitate of Calcium carbonate ($CaCO_3$) is formed.

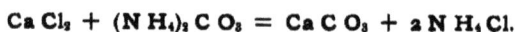

$$CaCl_2 + (NH_4)_2CO_3 = CaCO_3 + 2NH_4Cl.$$

186. A paste of plaster of Paris quickly " sets."

Experiment.—Mix some plaster of Paris (Calcium sulphate, $CaSO_4$) with water so as to make a stiff paste. Observe how quickly the paste now "sets" to a solid mass. (Make the paste on a piece of stiff paper.)

187. Alcohol expels Calcium sulphate from its solution in water.

Experiment.—Place a very small quantity of plaster of Paris in a test-tube. Add cold water and shake the tube, so as to favor the solution of the Calcium sulphate. Filter, and to the clear filtrate add its bulk of Alcohol; a white precipitate will appear. It is Calcium sulphate, which is slightly soluble in water, but is much less so in presence of Alcohol.

188. Quicklime and water unite with evolution of great heat.

Experiment.—Pulverize some *fresh* Quicklime; place a sufficient quantity of it in a casserole half-full of warm water; the Lime gradually unites with the water, forming Calcium hydroxide and affording great heat.

$$CaO + H_2O = CaO_2H_2.$$

CHAPTER VI. (*Continued.*)

THE METALLIC DYADS.—SECOND SECTION.

Mercury and Copper; Magnesium and Zinc.

OUTLINE OF THE SECOND SECTION.

Mercury.

Distribution.

Mercurous salts differ from Mercuric salts.

Metallic Mercury is precipitated by Copper and by Zinc.

Properties of Mercuric iodide and Mercuric sulphide.

Copper.

Distribution.

It conducts heat well; is difficult of fusion.

Metallic Copper is precipitated by metallic Iron.

Several tests for Copper.

Magnesium.

Distribution.

It burns in air, giving dazzling light.

It dissolves readily in acids.

Tests for Magnesium.

Zinc.

Distribution.

It burns readily; dissolves readily.

Tests for Zinc.

THE METALLIC DYADS.—SECOND SECTION.

189. The Second Section of the metallic dyads includes the following :

Name.	Symbol.	Ordinary condition.	Color.	Approximate Atomic weight.
Mercury,	**Hg,**	liquid,	white,	199.7
Copper,	**Cu,**	solid,	red,	63.2
Magnesium,	**Mg,**	solid,	white,	24.
Zinc,	**Zn,**	solid,	bluish-white,	64.9

Mercury, Hg. (Hydrargyrum.)

190. Distribution of Mercury.

Mercury is found, *in nature*, both as native Mercury and as Cinnabar (Mercuric sulphide, HgS).

In *the arts*, metallic Mercury is largely used, as is also Vermilion (HgS). Corrosive sublimate is Mercuric chloride ($HgCl_2$).

CAUTION.—Care must be taken to prevent metallic Mercury, or its solutions, from coming in contact with finger-rings or other jewelry. Mercury quickly alloys itself with Gold and with other metals, and produces stains upon them.

FIG. 43.—Mercury dissolving in Nitric acid.

191. Mercuric nitrate and its properties.

Experiment.—Dissolve, *completely*, a small globule of Mercury, by boiling it in concentrated Nitric acid. Mercuric nitrate is formed, **Hg (N O$_3$)$_2$.**

Divide the solution into two parts.

To the *first portion*, add a few drops of Hydrochloric acid; no precipitate should appear, because Mercuric chloride (**Hg Cl$_2$**) is soluble.

To the *second portion*, add Hydrochloric acid, and then Ammonium hydroxide; a white precipitate (Amido-mercuric chloride, **N H$_2$ Hg Cl**) appears.

192. Mercurous nitrate and its properties.

Experiment.—Dissolve, *only partially*, a globule of Mercury, by warming it in Nitric acid. Mercurous nitrate is formed, $Hg_2 (N O_3)_2$.

Dilute the solution, and then add a little diluted Hydrochloric acid; a white precipitate of Mercurous chloride ($Hg_2 Cl_2$) should appear. Filter, and to the white precipitate on the filter, add Ammonium hydroxide; a black precipitate should be formed—Amido-mercurous chloride (NH_2Hg_2Cl).

193. Metallic Copper precipitates metallic Mercury.

Experiment.—To a solution of Corrosive sublimate, add a few strips of Copper wire, which have been previously cleaned by immersion, first in Nitric acid and afterward in water; the wires soon become coated with a film of Mercury, which, if not already bright and silvery, may be made so by gentle rubbing with a cloth.

$$Cu_2 + Hg Cl_2 = Cu Hg + Cu Cl_2.$$

Dry the wires with filter-paper; place them in a narrow blow-pipe tube; heat them gently for a short time. The Mercury will volatilize from the Copper in vapors, which will condense to minute globules of liquid Mercury in the upper part of the tube.

194. Metallic Zinc precipitates metallic Mercury.

Experiment.—Try the same experiment as 193, only employ Zinc in place of Copper, and observe that the coating of Mercury renders the Zinc very brittle.

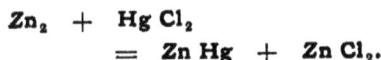

$$Zn_2 + Hg Cl_2$$
$$= Zn Hg + Zn Cl_2.$$

195. Mercuric iodide changes in color from salmon to scarlet.

FIG. 44.—Preparation of Mercuric Iodide.

Experiment.—To a solution of Corrosive sublimate, add, carefully, a solution of Potassium iodide. Mercuric iodide ($Hg I_2$) is formed.

$$Hg Cl_2 + 2 K I = Hg I_2 + 2 K Cl.$$

F

The Mercuric iodide goes through a series of delicate changes of color, from salmon to scarlet. Strangely enough, the precipitate is soluble in an *excess* either of Mercuric chloride or of Potassium iodide.

196. Mercuric sulphide, after some changes of color, becomes black.

Experiment.—To a solution of Corrosive sublimate, add some Sulphuretted-hydrogen as gas or dissolved in water. Precipitates varying from yellow to black may occur. They all contain more or less Mercuric sulphide (HgS).

$$HgCl_2 \ + \ H_2S \ = \ HgS \ + \ 2HCl.$$

Copper, Cu. (Cuprum.)

197. Distribution of Copper.

Copper occurs, *in nature*, in a great number of forms. Copper pyrites—a double sulphide of Copper and Iron—is the most important.

In *the arts*, next to the metal itself, the most important form is Cupric sulphate ($CuSO_4$), also called Blue vitriol.

198. Copper pyrites gives off Sulphur, when roasted.

Experiment.—Powder a few fragments of Copper pyrites and then heat them in a blow-pipe tube, and observe the Sulphur afforded.

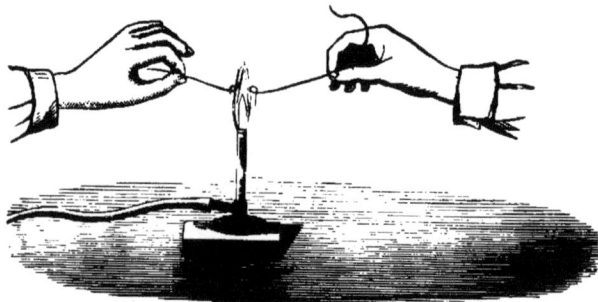

FIG. 45.—Observing the great difference between Copper and Platinum, as to their power of conducting heat.

199. Metallic copper is a good conductor of heat.

Experiment.—Hold in one hand a small Copper wire, and in the other hand a small Platinum wire; now simultaneously hold in a lamp-flame the disengaged ends of the wires, and observe the difference in the conducting powers of the metals.

200. Metallic Copper is not easily fusible before the blow-pipe.

Experiment.—Try to fuse a fragment of Copper wire before the blow-pipe, and upon a charcoal support. The metal is difficult of fusion.

201. Metallic Iron precipitates metallic Copper.

Experiment.—To a solution of Cupric sulphate, add a few drops of Hydrochloric acid. Now clean an Iron nail, or piece of Iron wire, by rubbing it with a cloth dipped in Hydrochloric acid. Immerse the cleaned Iron in the Copper solution, and allow the whole to stand until a considerable deposit of metallic Copper appears on the Iron.

$$Cu\,SO_4 \;+\; Fe \;=\; Fe\,SO_4 \;+\; Cu.$$

202. Copper dissolves slowly in Sulphuric acid, giving off fumes of Sulphur di-oxide (SO_2).

Experiment.—Add some concentrated Sulphuric acid to some strips of Copper wire. Now heat *with great care*. The Copper dissolves slowly, evolving the choking fumes of Sulphurous anhydride gas (SO_2).

$$Cu \;+\; 2H_2SO_4 \;=\; CuSO_4 \;+\; 2H_2O \;+\; SO_2.$$

203. Ammonium hydroxide is used as a test for Copper.

Experiment.—Dissolve a fragment of Cupric sulphate in water; filter, and to the filtrate add Ammonium hydroxide. If a sufficient quantity of the alkali is added, a clear and deep-blue solution is obtained.

204. Potassium ferro-cyanide is used as a test for Copper.

Experiment.—To a solution of Cupric sulphate add a few drops of solution of Potassium ferro-cyanide ($K_4\,Fe\,CN_6$). A rich brown precipitate of Cupric ferro-cyanide appears. (If it is desired to apply this test to an alkaline solution, Acetic acid—in quantity sufficient to neutralize the alkali—must first be added.)

$$2\,CuSO_4 \;+\; K_4FeCN_6 \;=\; Cu_2FeCN_6 \;+\; 2\,K_2SO_4.$$

205. Sulphuretted-hydrogen is used as a test for Copper.

FIG. 46.—Passing Sulphuretted-hydrogen gas into a solution of Copper.

Experiment.—To a very dilute solution of Cupric sulphate, add Sulphuretted-hydrogen as gas or its solution in water. A black precipitate of Cupric sulphide (**Cu S**) appears.

206. A process of testing Copper pyrites for Copper.

Experiment.—Grind a few fragments of Copper pyrites to a very fine powder. Place the powder in a test-tube, and after adding a little *Aqua-regia*, boil for a few minutes. Next, pour both liquid and sediment into a casserole containing water. Warm the solution, and filter it.

To the clear filtrate, add an excess of Ammonium hydroxide. This should precipitate the Iron and some other substances, but should dissolve the Copper.

Filter; and if the filtrate has a decided blue color, it may be considered as one test for the presence of Copper in the original ore. (See Experiment 204.)

Magnesium, Mg.

207. Distribution of Magnesium.

In *nature*, Magnesium is very abundant. Dolomites (Magnesian limestones) and Soapstones contain it.

In *the arts*, metallic Magnesium, Magnesium sulphate (Epsom salts, $MgSO_4$), and calcined Magnesia (MgO) are well known.

208. Magnesium wire or ribbon burns with dazzling light.

Experiment.—Hold a fragment of Magnesium wire in a pair of tweezers, and then light the Magnesium in the lamp-flame. A white ash of Magnesium oxide ($Mg\ O$) is produced by the combustion.

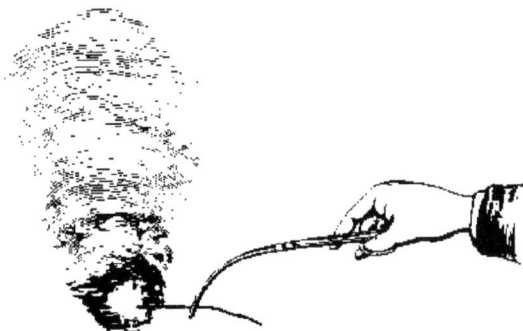

FIG. 47.—Magnesium ribbon burning, and producing Magnesium oxide ($Mg\ O$).

209. Magnesium dissolves easily in acids.

Experiments.—Dissolve one fragment of Magnesium wire in dilute Hydrochloric acid; it forms Magnesium chloride ($Mg\ Cl_2$).

Dissolve another fragment in dilute Nitric acid; it forms Magnesium nitrate, $Mg\ (N\ O_3)_2$.

Dissolve another fragment in dilute Sulphuric acid; it forms Magnesium sulphate ($Mg\ SO_4$).

$$Mg\ +\ H_2\ S\ O_4\ =\ Mg\ S\ O_4\ +\ H_2.$$

210. Ammonium salts have the remarkable power of preventing the precipitation of Magnesium compounds.

Experiment.—Dissolve some Magnesium sulphate in water, and divide the solution into two parts.

To the *first* part, add Potassium carbonate; a white precipitate appears.

$$Mg\ S\ O_4\ +\ K_2\ C\ O_3\ =\ Mg\ C\ O_3\ +\ K_2\ S\ O_4.$$

8

To the *second* part, add first Ammonium chloride solution in consider-able quantity, and then Potassium carbonate; the Ammonium chloride prevents the formation of a precipitate.

Zinc, Zn.

211. Distribution of Zinc.

In *nature*, many different ores of Zinc are known. Zinc carbonate ($ZnCO_3$) (Smithsonite) is an example.

In *the arts*, metallic Zinc is common; the name *galvanized Iron* is applied to Iron which has received a thin coating of metallic Zinc.

212. Metallic Zinc may be made to burn.

Experiment.—Heat a fragment of Zinc before the blow-pipe, on charcoal. The metal actually burns, forming Zinc oxide (**Zn O**).

213. Zinc dissolves easily in acids.

Experiments.—Dissolve one fragment of Zinc in dilute Hydrochloric acid; it forms Zinc chloride (**Zn Cl₂**).

Dissolve another fragment in dilute Nitric acid; it forms Zinc nitrate, **Zn (N O₃)₂.**

Dissolve another fragment in dilute Sulphuric acid; it forms Zinc sulphate (**Zn S O₄**). (See Experiments 43, 44, and 45.)

214. Zinc hydroxide is precipitated by Sodium hydroxide, but it re-dissolves.

Experiment.—Dissolve a small quantity of Zinc sulphate in water; add some solution of Sodium hydroxide; a white precipitate of Zinc hydrate (**Zn O₂ H₂**) appears. Now, add a considerable excess of Sodium hydroxide, and the Zinc hydroxide dissolves. Reserve this solution for Experiment 215.

215. Sulphide of Zinc is white.

Experiment.—To the alkaline solution, produced by Experiment 214, add Sulphuretted-hydrogen as gas or dissolved in water. A white precipitate of Zinc sulphide (**Zn S**) should appear.

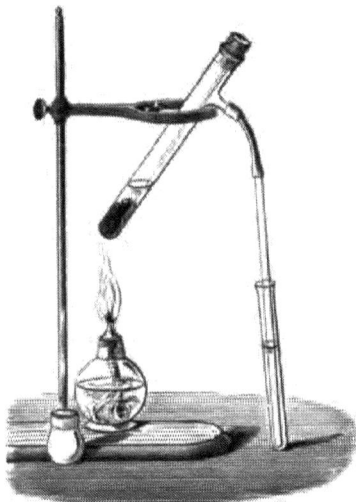

Fig. 48.—Passing Sulphuretted-hydrogen gas into a solution of Zinc

THE METALLIC DYADS.—THIRD SECTION.

Cobalt and Nickel ;
Iron, Manganese, Chromium ;
(and Aluminium).

OUTLINE OF THE THIRD SECTION.

Cobalt.

    Distribution.

    It forms a black Sulphide; it gives a blue color to Borax glass.

Nickel.

    Distribution.

    It is attracted by the magnet.

    It forms a black Sulphide; it gives a brown color to Borax glass

Iron.

    Distribution.

    Action of acids on wrought Iron and on cast Iron.

    Distinctive tests for Ferrous and Ferric salts.

    Ferrous compounds give bottle-green colors to Borax glass.

Manganese.

    Distribution.

    Tests for Manganese.

    Manganous compounds give a purple color to Borax glass.

Chromium.

>Distribution.
>
>Chromic acid is a powerful oxidizing agent.
>
>Chromium compounds give a green color to Borax glass.

(Aluminium.

>Distribution.
>
>The metal dissolves in alkalies.
>
>Properties of Alum.)

THE METALLIC DYADS.—THIRD SECTION.

216. The Third Section includes the metals of the following table. At different times all except Aluminium are dyads, tetrads, or hexads. Aluminium is usually a tetrad.

Name.	Symbol.	Ordinary condition.	Color.	Approximate Atomic weight.
Cobalt,	Co,	solid,	white,	58.9
Nickel,	Ni,	solid,	white,	57.9
Iron,	Fe,	solid,	white,	55.9
Manganese,	Mn,	solid,	white,	53.9
Chromium,	Cr,	solid,	gray,	52.
(Aluminium,	Al,	solid,	white,	27.

Cobalt, Co.

217. Distribution of Cobalt.

In *nature*, Cobalt is of somewhat rare occurrence; even in *the arts* it has but few uses. Its principal use is to impart a blue color to glass.

8 *

218. Cobalt forms a black Sulphide.

Experiment.—To a solution of Cobaltous nitrate, add first Ammonic hydrate, and then Sulphuretted-hydrogen as gas or dissolved in water. A black precipitate (Cobaltous sulphide, **Co S**) is formed.

Filter, and reserve the precipitate for the next experiment.

$$Co(NO_3)_2 + H_2S = CoS + 2HNO_3.$$

219. Cobalt compounds give a blue color to Borax glass.

Experiment.—Make a loop in a Platinum wire; dip the loop into powdered Borax, and then hold it in the lamp-flame. The Borax will lose its water of crystallization, with frothing (see 93). By heating sufficiently, a clear and colorless bead of Borax-glass is prepared.

Dip the bead into the black precipitate obtained by Experiment 218. Then fuse again in the lamp-flame. The bead should become dark-blue.

Nickel, Ni.

220. Distribution of Nickel.

Nickel occurs, *in nature*, in small quantities, but in a number of ores.

It is widely used in coinage and in Nickel-plating.

FIG. 49.—Metallic Nickel attracted by a magnet.

221. Metallic Nickel is magnetic.

Experiment.—Try a fragment of metallic Nickel with a magnet; it is attracted strongly.

222. Nickel forms a black Sulphide.

Experiment.—Dissolve, in water, a Double sulphate of Nickel and Ammonia ($Ni\ SO_4$ + $(N\ H_4)_2\ S\ O_4$. $6\ H_2\ O$); then add Ammonium hydroxide and Sulphuretted-hydrogen as gas, or dissolved in water. A black precipitate (Nickelous sulphide, $Ni\ S$) is formed. Reserve the precipitate for Experiment 223.

223. Nickel compounds give a brown color to Borax glass.

Experiment.—Fuse, into a fresh Borax bead, some of the black precipitate obtained by Experiment 222.

Compounds of Nickel make the bead *violet while hot*, and brown when cold.

Iron, Fe. (Ferrum.)

224. Distribution of Iron.

It is not probable that there exists *terrestrial* native Iron. But *meteorites* generally contain metallic Iron.

Many valuable oxides and other compounds of Iron are found in the earth as ores; but Iron pyrites (FeS_2), although abundant and widely diffused, is an ore that cannot be economically used for the manufacture of Iron.

In *the arts*, wrought Iron, Steel, and cast Iron are of immense importance. Ferrous sulphate ($FeSO_4$, also called Green vitriol and Copperas) is largely used.

225. The action of acids on wrought Iron.

Experiment.—Prepare three test-tubes; in them, boil portions of wrought Iron (carpet-tacks) in hydrochloric acid, in Nitric acid, and in Sulphuric acid respectively. Observe the differences in the action of the solvents.

$$Fe\ +\ 2\ H\ Cl\ =\ Fe\ Cl_2\ +\ H_2.$$

226. The action of acids on cast Iron.

Experiment.—Try an experiment like 225, only use cast Iron turnings in place of wrought Iron.

227. Iron forms Ferrous and Ferric salts.

Experiments.—(a) Prepare a *Ferrous* solution by dissolving some clean crystals of Ferrous sulphate ($Fe\,S\,O_4$) in water without heating.

(b) Prepare a *Ferric* solution by dissolving Ferrous sulphate in hot water, and then adding Nitric acid, and finally boiling the whole. Ferric sulphate, $Fe_2(S\,O_4)_3$, and Ferric nitrate, $Fe_2(N\,O_3)_6$, are formed. Dilute the solution for subsequent use.

228. The action of Ammonium hydroxide upon Iron solutions.

Experiments.—(a) Add Ammonium hydroxide to a portion of Ferrous solution. Ferrous hydrate is produced.

$$Fe\,S\,O_4 \;+\; 2\,N\,H_4\,O\,H \;=\; Fe\,O_2\,H_2 \;+\; (N\,H_4)_2\,S\,O_4.$$

(b) Add the same to a portion of Ferric solution. Ferric hydrate is produced.

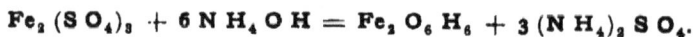

$$Fe_2\,(S\,O_4)_3 \;+\; 6\,N\,H_4\,O\,H \;=\; Fe_2\,O_6\,H_6 \;+\; 3\,(N\,H_4)_2\,S\,O_4.$$

229. Potassium ferro-cyanide is a test for *Ferric* salts only.

Fig. 50.—Producing Prussian-blue.

Experiments.—(a) Add solution of Potassium ferro-cyanide ($K_4\,Fe\,CN_6$) to a portion of Ferrous solution. A pale bluish-white precipitate is produced.

(b) Add the same to a portion of Ferric solution. Prussian-blue (Ferric ferro-cyanide) is produced.

230. Potassium ferri-cyanide is a test for *Ferrous* salts only.

Experiments.—(a) Add solution of Potassium ferri-cyanide ($K_3\,FeCN_6$) to a portion of Ferrous solution. A deep-blue precipitate, called Turnbull's blue, is produced.

(*b*) Add the same to a portion of Ferric solution; a reddish-brown coloration, but no precipitate, is produced.

231. Potassium sulpho-cyanate is a test for Ferric salts only.

Experiments.—(*a*) Add solution of Potassium sulpho-cyanate (KSCN) to a portion of Ferrous solution. No change of color is produced.

(*b*) Add the same to a portion of Ferric solution. A blood-red coloration, but no precipitate, is produced.

232. Metallic Iron is infusible, except at very high temperatures.

Experiment.—Try to fuse fragments of wrought Iron and of cast Iron, respectively, on charcoal.

233. Ferrous compounds give bottle-green colors to Borax glass.

Experiment.—Fuse, into a Borax bead, a minute fragment of Ferrous sulphate; colors are produced varying from yellow or bottle-green to dark-red, according to the conditions of the experiment.

FIG. 51.—Testing for Iron by a Borax bead

Manganese, Mn.

234. Distribution of Manganese.

Perhaps the most common and most useful *natural* form of Manganese is Pyrolusite (Manganese di-oxide, MnO_2).

In *the arts*, it is also used in the forms of Manganous sulphate ($MnSO_4$) and Potassium per-manganate ($K_2Mn_2O_8$).

235. Three tests for Manganese.

Experiments.—(*a*) To a solution of Manganous sulphate, add Ammonium hydroxide, and then Sulphuretted-hydrogen as gas, or dissolved in water; a flesh-colored precipitate of Manganous sulphide is formed (**MnS**). Upon exposure to air the sulphide becomes brown. Filter, and save the precipitate for Experiment (*b*).

(*b*) Dry the precipitate from Experiment (*a*), and then fuse it on a Platinum foil with a mixture of dry Potassium nitrate and Potassium carbonate; a green salt, Potassium manganate (**K₂ Mn O₄**) is formed. Proceed immediately to Experiment (*c*.)

(*c*) Place in a test-tube, half-full of water, the product of Experiment (*b*)—both the Platinum and the materials that are upon it. Warm the whole, gently, for a few moments, and then allow it to stand in quiet until the insoluble part subsides; the solution should have a reddish or purple color, owing to the formation of a small quantity of Potassium per-manganate. (This solution should not be filtered through paper; the latter decomposes the Potassium per-manganate sought.)

236. Potassium per-manganate ($K_2Mn_2O_8$) is a powerful oxidizing agent.

Experiment.—Dissolve some crystals of Potassium per-manganate in water. To this solution, add a solution of Ferrous sulphate and some Sulphuric acid.

The Potassium per-manganate oxidizes Ferrous sulphate to Ferric sulphate, itself becoming de-oxidized to Manganous sulphate, and the entire solution thereby becoming nearly or quite colorless.

237. Manganous compounds give a purple color to Borax glass.

Experiment.—Make a clear and colorless Borax bead on Platinum wire; into this bead fuse some Manganese di-oxide; the bead should acquire a violet or purple color.

Chromium, Cr.

238. Distribution of Chromium.

In *nature*, the most important form of Chromium is Chrome-iron ore (Ferroso-chromic oxide, $FeCr_2O_4$).

In *the arts,* Potassium di-chromate($K_2Cr_2O_7$) is its most important form.

239. Preparation of Chromic acid.

Experiment.— Make a concentrated solution of Potassium di-chromate by boiling some of the powdered salt in a small quantity of water; filter while hot; to the filtrate, add *very carefully* about its bulk of concentrated Sulphuric acid. Allow the solution to cool, when dark-red crystals of Chromic acid should appear.

240. Chromic acid is a powerful oxidizing agent.

Experiment.—Make a solution of a small quantity of Potassium di-chromate; to it, add a small quantity of Sulphuric acid; now add Alcohol, *drop by drop, with great care.*

The red color, due to Chromic acid (see Experiment 239), quickly changes to a beautiful green. This change is due to the oxidizing power of the Chromic acid, and the *reducing* action of the Alcohol. The green substance is Chromic sulphate, $Cr_2(SO_4)_3$.

(Proceed to use this product for Experiment 241.)

241. Ammonium hydroxide is a test for Chromic salts, but not for Chromates.

Experiments.—(a) Carefully evaporate, to one-half its bulk, the solution formed by Experiment 240. Dilute the residue with water. To the clear green solution, add Ammonium hydroxide. A dull-green precipitate of Chromic hydroxide ($Cr_2O_6H_6$) appears. (The color is recognized after boiling.)

$$Cr_2(SO_4)_3 + 6\,NH_4OH = Cr_2O_6H_6 + 3\,(NH_4)_2SO_4.$$

(b) To a clear solution of Potassium di-chromate, add Ammonium hydroxide. No visible chemical change takes place.

242. Potassium di-chromate is an oxidizing agent.

Experiment.—Make a solution of Potassium di-chromate in water. Pour a few drops of the solution upon a clean filter-paper. Carefully dry the filter-paper over the lamp. When quite dry, apply a burning match to the edge of the paper. It burns steadily, but without flame, and leaves a green tea-like ash. The combustion is *assisted* by the Oxygen of the

Potassium di-chromate, but it is *retarded* by the other constituents of the compound. The green color of the ash is due to the formation of some compound of Chromium.

243. Chromium compounds give a green color to Borax glass.

Experiment.—Make a clear and colorless Borax bead on a platinum wire; into it, fuse a minute crystal of Potassium di-chromate; an emerald-green color is produced.

Aluminium, Al.

244. Distribution of Aluminium.

Compounds of Aluminium are among the three or four most abundant mineral materials in the earth. Clay, a complex silicate of Aluminium, is an example.

Alum (Double sulphate of Aluminium and ammonium), $Al_2(SO_4)_3.(NH_4)_2SO_4.24H_2O$, is very largely used in dyeing. The metal Aluminium is slightly used.

245. Alum easily forms crystals.

Experiment.—Pulverize some Alum. Dissolve a considerable quantity of it, by boiling it in a small quantity of water. Filter, and allow the filtrate to stand at rest for twenty-four hours. The Alum should form crystals upon cooling.

246. Ammonia may be detected in Alum.

Experiment.—There are many kinds of Alum. Dissolve a fragment of ordinary Alum in water. Now test the solution for Ammonia-gas, as described in Experiment 99. The Ammonia-gas will probably be discovered, since Ammonia-alum is that frequently used at present.

247. Aluminic hydroxide dissolves in alkalies.

Experiment.—Dissolve a small crystal of Alum in water; add a few drops of solution of Sodium hydroxide, and boil; a flaky precipitate of Aluminic hydroxide ($Al_2O_3H_6$) appears. Now add considerably more Sodium hydroxide, and boil again; the precipitate is soluble in a considerable excess of the alkali.

FIG. 52.—Crystals of Alum.

248. Even metallic Aluminium dissolves in alkali.

Experiment.—Boil a fragment of Aluminium in a solution of Sodium hydroxide; it dissolves, evolving Hydrogen gas and producing Sodium aluminate ($Na_2 Al_2 O_4$).

G

CHAPTER VII.

THE METALLIC TRIADS.

Bismuth and Gold.

OUTLINE OF THE CHAPTER.

Bismuth.

    Distribution.

    The metal is brittle.

    Its solutions give white precipitates by mixture with water.

Gold.

    Distribution.

    Test for Gold, by producing Purple of Cassius.

THE METALLIC TRIADS.

249. The principal metallic triads are the following:

Name.	Symbol.	Ordinary condition.	Color.	Approximate Atomic weight.
Bismuth,	**Bi**,	solid,	reddish-white,	207.5
Gold,	**Au**,	solid,	yellow,	196.2

Bismuth, Bi.

250. Distribution of Bismuth.

Bismuth and its compounds are comparatively rare *in nature* and in *the arts*.

251. The metal is brittle.

Experiment.—Pulverize a fragment of metallic Bismuth; observe that it is brittle, while most metals are malleable.

252. Bismuth dissolves in Nitric acid.

Experiment.—Dissolve the powder, from Experiment 251, by warming it in dilute Nitric acid. Evaporate the solution to a few drops, and then pour it into a beaker nearly full of cold water; a white precipitate appears (Bismuthyl nitrate, [$Bi\,O$] $N\,O_3$).

253. Bismuthous sulphide is black.

Experiment.—Dissolve Bismuthyl nitrate ($Bi\,O\,N\,O_3$, Basic Nitrate of Bismuth) in Hydrochloric acid; then pour the solution into a beaker half-full of cold water; a white precipitate of Bismuthyl chloride ($Bi\,O\,Cl$) appears. Add now Sulphuretted-hydrogen as gas, or dissolved in water, when a black precipitate of Bismuthous sulphide ($Bi_2\,S_3$) will be formed. Filter, and reserve the precipitate.

FIG. 53.—Bismuth dissolving in Nitric acid.

254. Metallic Bismuth may be produced from its compounds.

Experiment.—Remove from the filter the precipitate obtained by Experiment 253, and then fuse it, on charcoal, with Potassium carbonate. A metallic globule of Bismuth should be obtained. Place it in a mortar, and ascertain whether it is brittle or not.

Gold, Au. (Aurum.)

255. Distribution of Gold.

In *nature*, Gold is generally found in the metallic or uncombined state. Although it is very widely distributed, it is nowhere abundant.

In *the arts*, although employed for almost numberless purposes, it is almost invariably used in the metallic state.

256. Gold is detected by its affording *Purple of Cassius.*

Experiment. — Dissolve some Gold-leaf as described in Experiment 112. Also, prepare a test-liquid by adding a solution of Stannous chloride to a solution of Ferric chloride. (Ferric chloride may be produced by dissolving a few fragments of fine Iron wire in Hydrochloric acid, then adding a few drops of Nitric acid, and boiling for a minute.) Now add a few drops of the Gold solution to the test-liquid. A precipitate called *Purple of Cassius* appears; it varies in color, being brown, blue, purple, or black, according to the conditions under which it is produced.

FIG. 54. — Detecting Gold by producing Purple of Cassius.

CHAPTER VIII.

THE METALLIC TETRAD.

Platinum.

OUTLINE OF THE CHAPTER.

Platinum.

It occludes gases.

It dissolves in Aqua-regia.

THE METALLIC TETRAD.

257. The principal metallic tetrad is the following.

Name.	Symbol.	Ordinary condition.	Color.	Approximate Atomic weight.
Platinum,	Pt,	solid,	white,	194.4

258. Distribution of Platinum.

In *nature*, Platinum occurs in the metallic state, in the condition of an alloy with certain other metals.

In *the arts*, it has valuable applications; it is generally used in the metallic form.

259. Platinum re-lights an extinguished gas-jet.

Experiment.—Heat a piece of tolerably clean Platinum foil in a Bunsen lamp flame. Now stop the gas, and soon let it flow anew against the Platinum. The metal quickly becomes red-hot, and often re-lights the gas. The Platinum absorbs, or occludes, upon its surfaces, both the coal-gas and the Oxygen of the air; the two substances are thus brought within the range of chemical affinity, and so they unite, affording heat and light.

When illuminating-gas is not at hand, the experiment may be performed as follows: Boil some water in a casserole or a beaker. Move the lamp to a safe distance. In the hot water, place a small beaker containing alcohol; the upper part of the beaker soon fills with vapor of alcohol. Now make a coil by winding a Platinum wire, in a close spiral, around a lead-pencil. Heat the spiral in a lamp-flame; then suspend it in the alcohol vapors previously described. The wire should continue to glow, by reason of a slow combustion of the alcohol vapors.

FIG. 55.—Metallic Platinum producing a flameless combustion of Alcohol vapor.

260. Platinum dissolves in Aqua-regia.

Experiment.—Dissolve a small fragment of Platinum wire in *Aqua-regia*. Evaporate the solution nearly to dryness; dilute this product slightly with water. Add a solution of Ammonium chloride ($N H_4 Cl$). A yellow crystalline precipitate of Ammonio-platinic chloride appears, $(N H_4)_2 Pt Cl_6$. It proves the presence of Platinum in the solution.

Appendix.

LIST OF SUPPLIES

Needed for the Performance of the Experiments described in "The Young Chemist."*

1. Alum, $(N H_4)_2 S O_4, Al_2 (S O_4)_3. 6 H_2 O.$
2. Aluminium, $Al.$
3. Ammonium carbonate, $(NH_4)_2 C O_3.$
4. — chloride, $N H_4 Cl.$
5. Antimony, $Sb.$
6. Argentic nitrate, $Ag N O_3.$
7. Arsenious oxide, $As_2 O_3.$

8. Barium nitrate, $Ba (N O_3)_2.$
9. — chloride, $Ba Cl_2. 2 H_2 O.$
10. Beeswax.
11. Bismuth, $Bi.$
12. Bismuthyl nitrate, $Bi O N O_3.$
13. Bleaching powder, $[Ca O_2 Cl_2 + Ca Cl_2 + Ca O_2 H_2].$
14. Borax, $Na_2 B_4 O_7.$

15. Calcium carbonate (marble), $Ca C O_3.$
16. — sulphate, $Ca S O_4.$
17. Carbon di-sulphide, $C S_2.$
— Charcoal, $C.$
18. — animal, $C.$
19. Cobaltous nitrate, $Co (NO_3)_2.$
20. Copper (wire), $Cu.$
21. — pyrites, $Cu_2 S + Fe_2 S_3.$
22. — sulphate, $Cu S O_4.$

23. Fluor-spar, $Ca F_2.$

* Complete sets of apparatus and supplies may be obtained of Eimer & Amend, cor. 3d Ave. and 18th St., New York.

24. Gold-leaf,   Au.

25. Indigo.
26. Iodine,   I.
27. Iron pyrites,   Fe S$_2$.
28. — sulphate,   Fe S O$_4$.
29. — sulphide,   Fe S.
30. — (turnings),   Fe $+$ C.
31. — (wire),   Fe.

32. Lead (sheet),   Pb.
33. — acetate,   Pb O$_2$ (C$_2$ H$_3$ O)$_x$
34. — nitrate,   Pb (N O$_3$)$_2$.
— Lime (quick),   Ca O.
35. Lithium carbonate,   Li$_2$ C O$_3$
36. Litmus-blocks.
37. — paper.

38. Magnesium,   Mg.
39. — sulphate,   Mg S O$_4$. 7 H$_2$ O.
40. Manganese di-oxide,   Mn O$_2$.
41. — sulphate,   Mn S O$_4$.
42. Mercury (metallic),   Hg.
43. Mercuric chloride,   Hg Cl$_2$.
44. — oxide,   Hg O.

— Nickel coin,   (Nickel, Copper, Zinc).
45. — metallic,   Ni.
46. — double sulphate of, } Ni S O$_4$, (N H$_4$)$_2$ S O$_4$. 6 H$_2$ O
     and ammonia,   }

47. Oxalic acid,   H$_2$ C$_2$ O$_4$. 2 H$_2$ O.

48. Paraffin,   C$_n$H$_{2n+2}$.
49. Phosphorus,   P.
50. Potassium,   K.
51. Potassium bromide,   K Br.
52. — iodide,   K I.
53. — carbonate,   K$_2$ C O$_3$.
54. — chlorate,   K Cl O$_3$.
55. — di-chromate,   K$_2$ Cr$_2$ O$_7$.

56. Potassium ferro-cyanide,　$K_4 Fe CN_6$
57. — ferri-cyanide,　$K_6 Fe_2 CN_{12}.$
58. — nitrate,　$K N O_3.$
59. — per-manganate,　$K_2 Mn_2 O_8.$
60. — sulpho-cyanate,　$K S CN.$

61. Quill.

62. Sand,　$Si O_2.$
63. Shellac.
— Silver coin,　(Silver, Copper.)
64. Sodium,　$Na.$
65. Sodium chloride,　$Na Cl.$
66. — hydroxide,　$Na O H \ + \ Aq.$
67. — silicate,　$Na_2 Si O_3 + Aq.$
68. Stannous chloride,　$Sn Cl_2, 2 H_2 O.$
— Starch,　$C_6 H_{10} O_5.$
69. Strontium nitrate,　$Sr (N O_3)_2.$
— Sugar (cane),　$C_{12} H_{22} O_{11}.$
70. Sulphur,　$S.$
— Sulphuretted-hydrogen
　　water,　$H_2 S.$

71. Tartar emetic,　$K Sb O H_2, O_4, (C_4 H_2 O_2). \ + \ Aq$
72. Tinfoil,　$Sn.$
— Turpentine,　$C_{10} H_{16}.$

73. Zinc sulphate,　$Zn S O_4. 7 H_2 O.$
74. — metallic,　$Zn.$

Acid, Acetic,　$H O (C_2 H_3 O).$
— Hydrochloric,　$H Cl.$
— Nitric,　$H N O_3.$
— Sulphuric,　$H_2 S O_4.$
Alcohol, Ethylic,　$C_2 H_5 O, H.$
Ammonium hydroxide,　$N H_4 O H.$

LIST OF APPARATUS

3 Beakers (small).
3 Blocks (of wood; 3 × 3 × 1 inches).
1 Blow-pipe (brass).
1 Casserole (six ounce).
  Corks.
1 Cork (fitted with glass-jet).
1 Crayon.
1 Crucible (porcelain).
25 Filter-papers.
1 Flask (plain, eight ounce).
1 — (side-neck, two ounce).
1 Funnel (two inch).
2 feet Glass tubing (quill size).
1 Glass rod.
1 Jar (species).
1 Lamp, for alcohol (or for gas; with tube).
1 Lamp-chimney.
1 Lead cup.
1 Lead post (and rubber-ring).
1 Magnet.
1 Mortar.
1 Platinum foil (one inch square).
1 — wire (three inches).
1 Retort (two ounce, tubulated).
1 Rubber tube.
12 Test-tubes (plain, five-inch).
2 — — (— eight inch).
1 — — (side-neck, five inch).
1 — — (hard glass, eight inch).
1 — — (side-neck, eight inch).
6 — — (small, for blow-pipe).
1 Test-tube rack (for twelve tubes).
1 Taper.
1 pair Tweezers.
1 Tripod (or Triangle, or Lamp-stand).
1 Watch-crystal.

INDEX.

ACID, arsenic, 21.
  arsenious, 21.
  boric, 46, 47, 48.
  carbonic, 63, 64.
  chloric, 19.
  chlorous, 19.
  chromic, 95.
  hydrochloric, 32.
    preparation of, 32.
    tests for, 33.
  hydrofluoric, 29, 30.
  hypochlorous, 19.
  meta-stannic, 66.
  muriatic, 16.
  nitric, 16, 46, 51, 52.
    preparation of, 54.
    tests for, 53, 54.
  oxalic, 62.
  perchloric, 19.
  phosphoric, 46, 55.
  silicic, 65.
  sulphuric, 28, 29, 37, 43–45.
Acids, 19.
  haloid, 18.
Alcohol, 62.
Alum, 96.
Aluminic silicate, 96.
Aluminium 89, 96.
Amidogen, 48.
Amido-mercuric chloride, 80.
  -mercurous chloride, 81.
Ammonia-gas, 46, 48–50.
Ammonium chloride, 49, 50.
  cyanide, 17.
Ammonium, 48.
  salts, 50.
Anhydrides, 18.
Antimonic oxide, 57.
Antimonious sulphide, 47, 58.
Antimony, 47, 57.
  and potassium, tartrate of, 57.
  tetroxide, 57.
Aqua-regia, 53.

Argentic bromide, 34.
  chloride, 32.
  iodide, 36.
  nitrate, 32.
Argentum, 68.
Arsenic, 47, 55.
  white, 55, 56.
Arsenious oxide, 55.
  sulphide, 47, 57.
Atomic weights (table), 13.
Aurum, 100.

BARIUM chloride, 75
  nitrate, 75.
  sulphate, 75.
Barium, 72, 75.
Base, 21.
Binaries, 17.
Black-lead, 60.
Bleaching with sulphur, 43.
  powder, 30, 31, 77.
Blue vitriol, 82.
Bone, 54.
Borax, 47, 48.
Boron, 46, 47.
Brimstone, 42.
Bromine, 25, 26, 33.
  preparation of, 34.

CALCIUM carbonate, 77, 78.
  chloride, 77, 78.
  fluoride, 29.
  phosphate, 54.
  sulphate, 78.
Calcium, 72.
Carbon, 59–61.
  di-oxide, 41, 63, 64.
  mon-oxide, 62.
Caustic soda, 71.
Chlorates, 42.
Chlorine, 25, 26, 30.
  bleaching with, 31.
Chlorine, preparation of, 31.

Chrome-iron ore, 94.
Chromium, 89, 94.
Chrome-yellow, 74.
Cinnabar, 80.
Clay, 96.
Coal, 60.
Cobalt, 88, 89.
Compound substances, 12.
  names of, 15.
Copper, 79, 82, 83.
  pyrites, 82, 84.
Copperas, 91.
Corrosive sublimate, 80–82.
Cupric ferro-cyanide, 83.
  sulphate, 82, 83.
  sulphide, 84.
Cuprum, 82.
Cyanogen, 17.

Diamond, 60.
Dolomites, 84.
Dyads, metallic, 72, 79, 88.
  non-metallic, 37.

Elements, 11–14.
  names of, 14.
  symbols of, 14.
  table of, 13.
Epsom salts, 84.
Ethylene, 61, 62.
Etching with Hydrofluoric acid, 30.

Ferric compounds, 92, 93.
Ferroso-chromic oxide, 94.
Ferrous compounds, 92, 93.
  nitro-sulphate, 53.
  sulphide, 45.
Ferrum, 91.
Fluorine, 25, 26, 29.
Fluor-spar, 29, 30.

Galena, 16, 73.
Glass, 65.
Gold, 100.
  injured by mercury, 80.
Graphite, 60.
Green-fire, 76.
Gum-shellac, 76, 77.

Hartshorn, spirits of, 49.
Heavy-spar, 75.
Hydrargyrum, 80.
Hydrogen, 25–29.

Hydrogen, compounds of, with nitrogen, 46, 48.
Hydro-potassium sulphate, 21, 54.
  -sodic sulphate, 32.

Illuminating gas, 61.
Indigo, 45, 51, 61.
Iodine, 25, 26, 35.
  preparation of, 34.
Iron (see Ferrous, etc.), 41, 88, 91.
  galvanized, 86.
  magnetic oxide of, 41.
  pyrites, 42, 91.

Kalium, 69.

Laughing-gas, 50.
Lead, 72–75.
  carbonate of, 73.
  chloride of, 73.
  chromate of, 74.
  iodide of, 74.
  nitrate of, 73.
  sulphide of, 74.
Lead-post, 8.
Lime, 77.
  chloride of, 30, 31.
  quick, 78.
  -water, 64.
Limestone, 77.
Lithium, 67, 68, 71.
Litmus, 43.

Magnesia, 84.
Magnesic sulphate, 84.
Magnesium, 79, 84, 85.
Manganese, 88, 93.
  di-oxide, 31, 34, 39, 40.
Manganous chloride, 34.
  sulphate, 93.
Marble, 63, 77.
Marsh-gas, 16, 61.
Mercuric chloride, 17, 80.
  iodide, 81.
  nitrate, 80.
  oxide, 39.
  sulphide, 80, 82.
Mercurous chloride, 17, 81.
  nitrate, 81.
Mercury, 79–81.
  red oxide of, 39.
Metallic dyads, 72, 79, 88.
  monads, 67.
  tetrad, 101.

Metallic triads, 98.
Meteorites, 91.
Methyl hydride, 61.
Mixture (defined), 12.
Monad, definition of, 25.
Monads, metallic, 67.
  non-metallic, 25.

NATRIUM, 71.
Nickel, 88, 90.
Nitrates, 41, 48.
Nitric anhydride, 50.
Nitrogen, 46–48.
  and oxygen, compounds of, 46, 50.
  compounds of, with hydrogen, 46, 48.
  di-oxide, 50, 51.
  pentoxide, 50.
  protoxide, 50.
  tetroxide, 50, 51.
  tri-oxide, 50.
Nitrous anhydride, 50.
Nomenclature, 11, 14.
Notation, 11, 14.

OLEFIANT gas, 61, 62.
Oxygen, 37–42.
  and nitrogen, compounds of, 46, 50.

PAPER, 44.
Paraffin, 61.
Phosphoric anhydride, 55.
  oxide, 55.
Phosphorus, 46, 47, 54.
Photographic "proof," 32.
Plaster of Paris, 78.
Platinum, 29, 101.
Plumbago, 60.
Plumbic acetate, 45.
  carbonate, 73.
  chloride, 73.
  chromate, 74.
  iodide, 74.
  nitrate, 73.
  sulphate, 43.
  sulphide, 16, 45, 73, 74.
Plumbum, 73.
Potassium arsenate, 21.
  bromide, 33.
  carbonate, 63, 69, 70.
  chlorate, 20, 39, 40, 42, 69, 70, 76, 77.

Potassium chlorite, 20.
  di-chromate, 69, 70, 71, 95.
  ferro-cyanide, 83.
  ferri-cyanide, 92.
  hydrate, 27, 69.
  hypochlorite, 20.
  iodide, 34–36, 81.
  nitrate, 41, 48, 60, 69, 70.
  oxide, 27.
  perchlorate, 20.
  per-manganate, 93, 94.
  sulpho-arsenate, 21.
  sulpho-cyanate, 93.
Potassium, 26, 27, 67–69.
Prefixes, 18.
Priestley, Dr., 39.
Prussian blue, 92.
Purple of Cassius, 100.
Pyrolusite, 93.

QUARTZ, 64.
Quicklime, 78.
Quill, 51.

RADICLES, compound, 17.
Red-fire, 77.
Rock crystal, 65.
  salt, 71.

SALT, common, 30.
Salts, 20.
  acid, 21.
  basic, 21.
  haloid, 18.
  normal, 21.
  sulphur, 21.
Saltpetre, 48.
Sand, 64, 65.
Selenium, 37.
Sheet-tin, 66.
Silicates, 65.
Silicic fluoride, 30.
  oxide, 30, 65.
Silicon, 59, 60, 64.
Silver, 67, 68.
Smithsonite, 86.
Snow-crystals, 26.
Soap-stone, 84.
Soda-ash, 71.
Sodium carbonate, 71.
  chloride, 30, 71.
  hydroxide, 27, 71.
  oxide, 27.

10

Sodium silicate, 65.
  stannate, 66.
Sodium, 27, 28, 67, 68, 71.
Stannic oxide, 66.
Stannous chloride, 66.
  sulphide, 66.
Stannum, 66.
Starch, 35, 36, 44, 60, 62.
  iodide of, 36.
Stibium, 57.
Stibnite, 57.
Strontium chloride, 76.
  nitrate, 76, 77.
  sulphate, 76.
Sugar, 44, 62.
Sulphur, 37, 38, 41, 42.
Sulphuretted-hydrogen, 37, 45.
Sulphurous anhydride, 83.
  di-oxide, 41, 43, 83.
Sunlight on silver salts, 32.
Supports for apparatus, 9.
Symbols, literal, 14, 16.
  graphic, 15, 16, 22, 23.
  glyptic, 15, 17.
  of acids 22, 23.
  of compounds, 16, 17, 22, 23.
  of salts, 22, 23.

TARTAR-emetic, 57.
Teachers, hints to, 7.
Tellurium, 37.
Ternary compounds, 91.
Tetrads, metallic, 101.
  non-metallic, 59.
Tin, 59, 60. 66.
Tin crystals, 66.
Titanium, 59, 60.
Triads, metallic, 98.
  non-metallic, 46.
Turnbull's blue, 92.
Turpentine, 61.

VITRIOL, blue, 82.
  green, 91.
  oil of, 30.

WHITE arsenic, 55, 56.
  lead, 73.
Wood, 60.

ZINC, 28, 79, 80, 86.
  carbonate of, 86.
  sulphate, 29
  sulphide, 87.

Milton Keynes UK
Ingram Content Group UK Ltd.
UKHW020623071223
433866UK00005B/138